ENDANGERED ANIMALS

ENDANGERED ANIMALS

TOM JACKSON

amber
BOOKS

This pocket edition first published in 2026

First published in 2022

Published by
Amber Books Ltd
United House
London N7 9DP
United Kingdom
www.amberbooks.co.uk
Facebook: amberbooks
YouTube: amberbooksltd
Instagram: amberbooksltd
X(Twitter): @amberbooks

ISBN: 978-1-83886-678-5

Editorial, design and picture research: Amber Books Ltd

Printed in China

Contents

Introduction

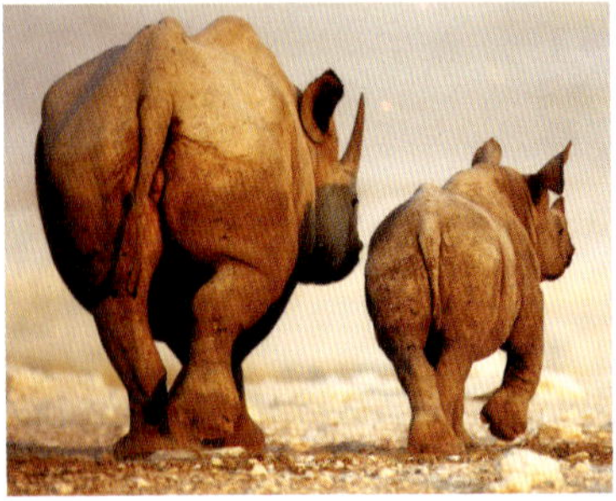

The natural world is facing a crisis. A total of 680 vertebrates have been driven to extinction in the last 500 years, including the dodo, the marsupial tiger, and Steller's sea cow. Yet that could just be the beginning. Three-quarters of all land habitats have been degraded by humans, if not completely altered. Even the oceans, which make up 99 percent of the biosphere where life survives, show many consequences of human activity. It is estimated that about one million species of animal are now threatened with extinction, mainly insects and mostly completely unknown to science.

Extinction is a natural process, but naturalists calculate that the current rate of extinction is 1,000 times faster than normal. That scale of threat suggests that we are in a mass extinction of the kind last seen when the dinosaurs were wiped out.

The plight of the natural world is well understood thanks to highly publicized reports of the perils faced by iconic animals, such as the panda, rhino or tiger. As this book shows, there are plenty more beautiful creatures in grave danger.

ABOVE:
Targeted for their horns, black rhinos only survive thanks to conservation activities.

OPPOSITE:
The orang-utang is the rarest great ape. Their numbers are still in decline as their jungle home is steadily destroyed by fire and farmers.

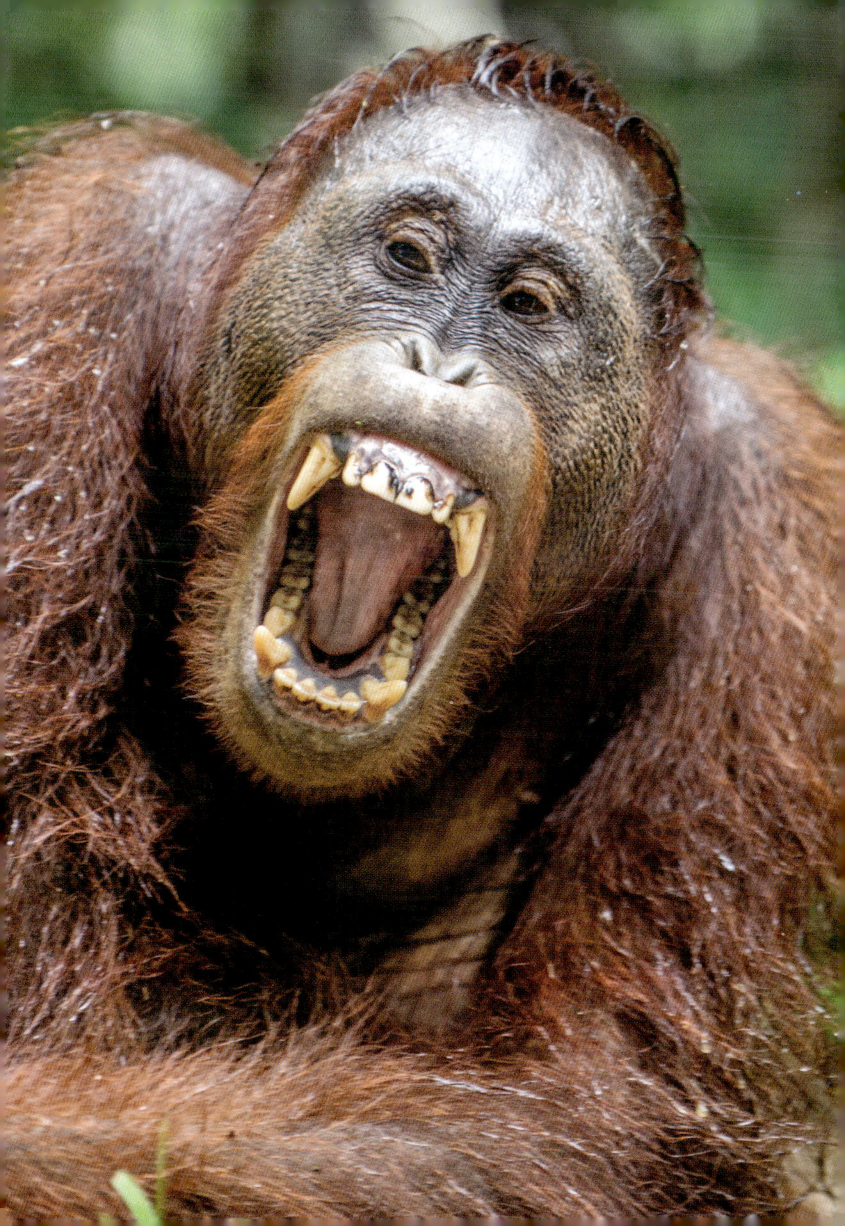

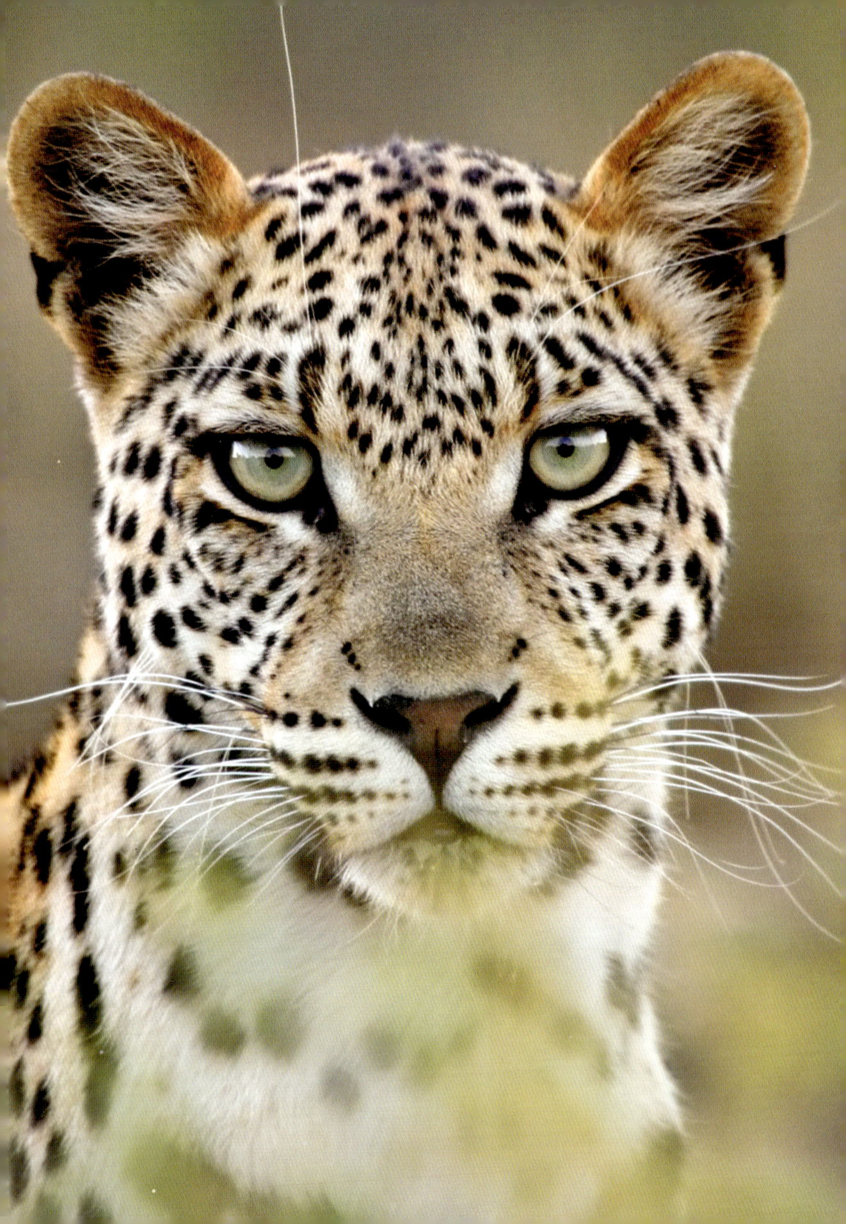

Africa

The second-largest of the continents, Africa has many iconic landscapes and habitats that are home to many of the most famous – and endangered – of the world's animals. These places include the Sahara, the world's largest desert, which is home to the addax and oryx. Thanks to over-zealous hunting, these are now two of the rarest antelopes species of all. To the south lie lush jungles, where many species of primate, including our close relatives the chimps and gorillas, are under threat from habitat destruction and even hunters looking for bush meat, which is still sadly in demand.

Perhaps the most famous African animals of all are found in the savannah, a region of hot grasslands and dry woodlands to the east and south of the continent. These include majestic giants like the African elephants, the rhinos and giraffes, plus the big cats that rule over these lands. These mighty beasts are struggling to survive as they come up against interference from humans. Firstly, farmers are turning over the grassland to farming and they kill wild animals that threaten their livelihoods. More significantly, there is the threat from poachers who are paid vast sums to kill Africa's great animals for use in spurious medicines, and as trophies.

OPPOSITE:

Leopard

Africa's second-largest big cat species, the leopard is also an Asian species. It is vulnerable across its range as it runs out of places – and prey – to hunt.

ALL PHOTOGRAPHS:

Leopard

With Africa's farmland set to swell by 20 per cent over the next 25 years, an accommodation needs to be made with the leopard. These lone predators are most active at night. By day, they hang out in trees, but the lack of hunting space is forcing the big cats to push deeper into human territory. Found mostly in Sub-Saharan Africa, this spotted killer lives as far east as Korea and Java. A melanistic version in India is known as the black panther. Asian leopard subspecies are at great risk from poaching and most survive in populations of a few hundred.

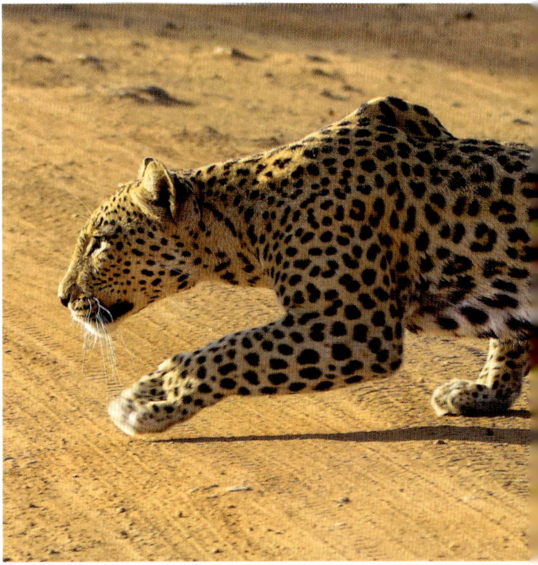

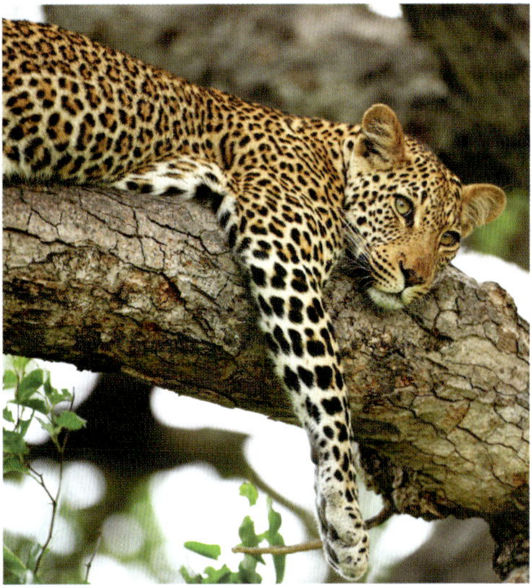

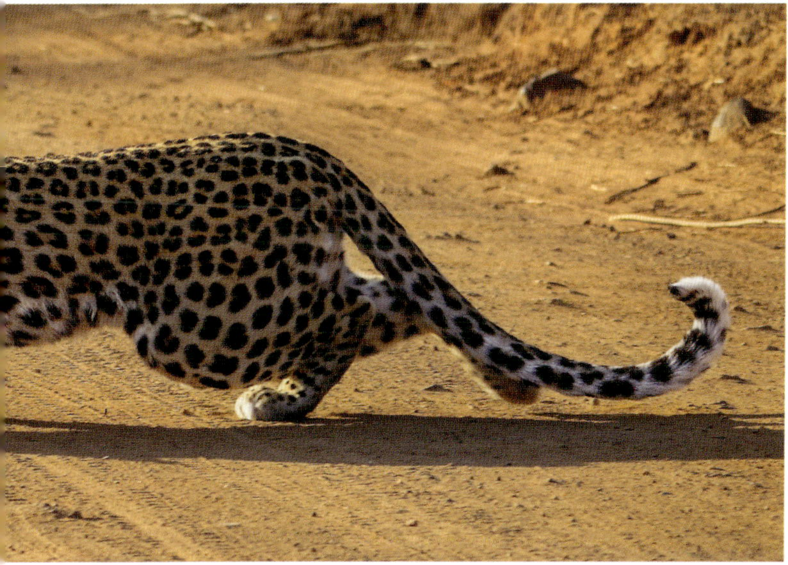

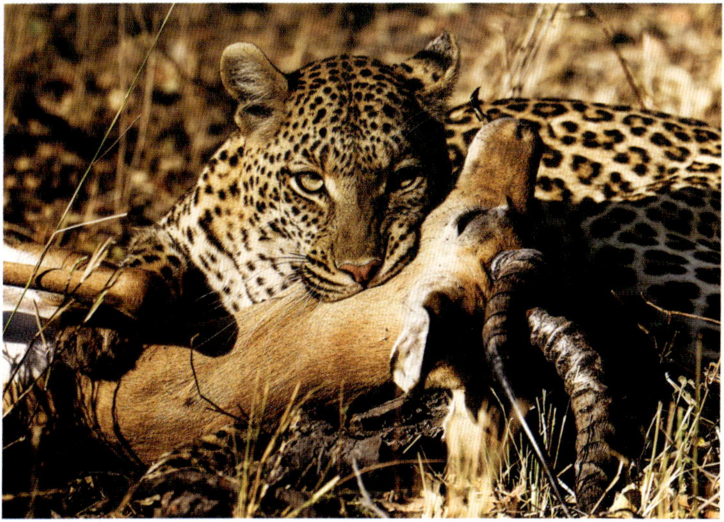

Addax

This large antelope is supremely adapted to life in the scorching aridity of the Sahara and its surroundings. Its pale hide reflects light and heat to keep it cool in the desert heat. However, fewer than 100 of them survive in the wild.

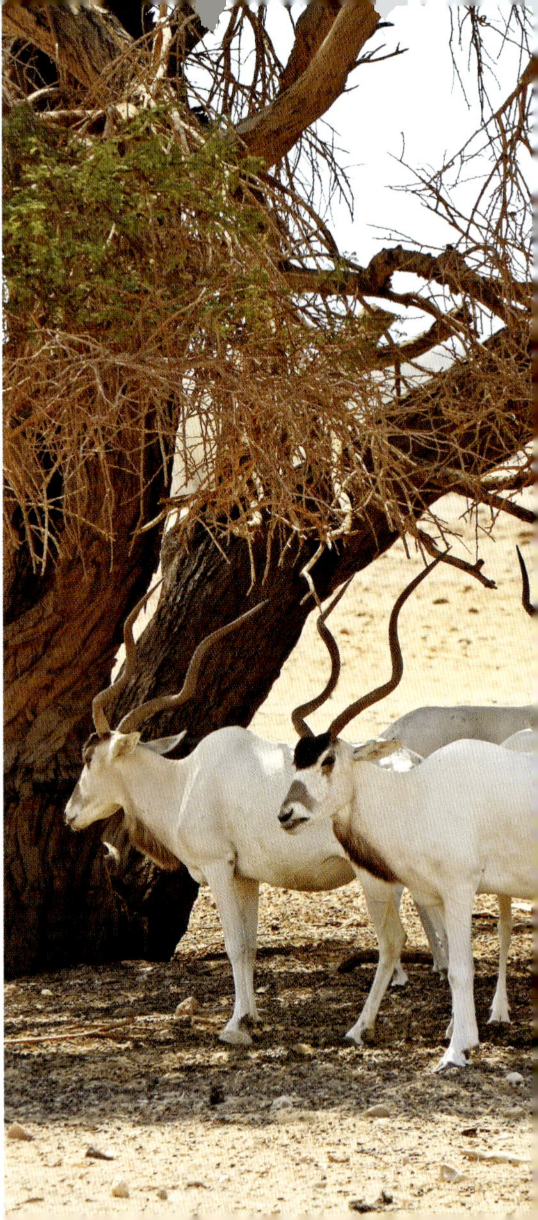

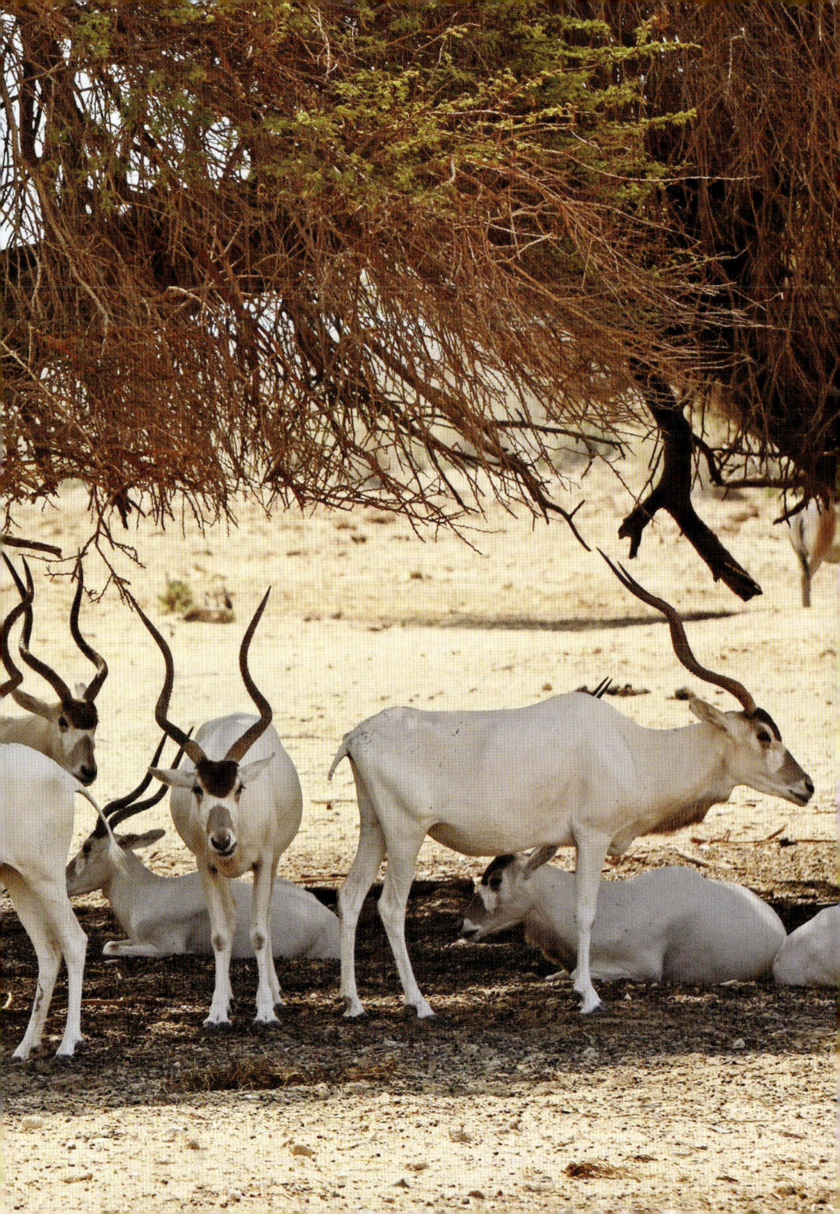

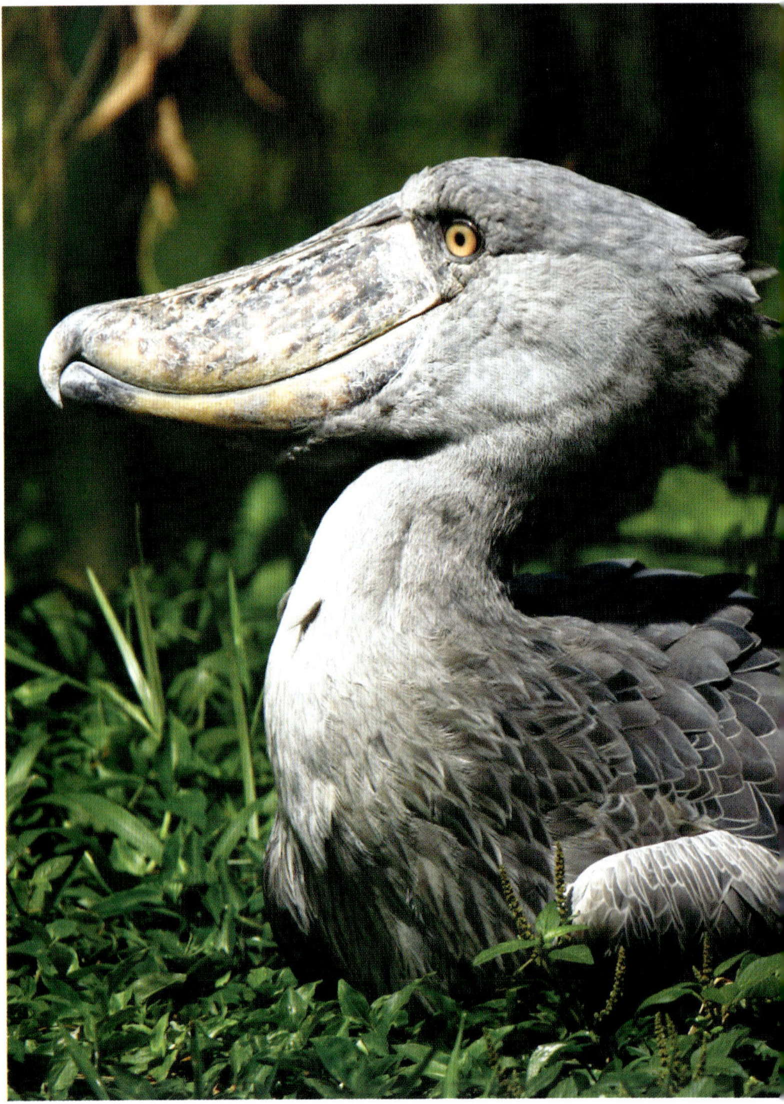

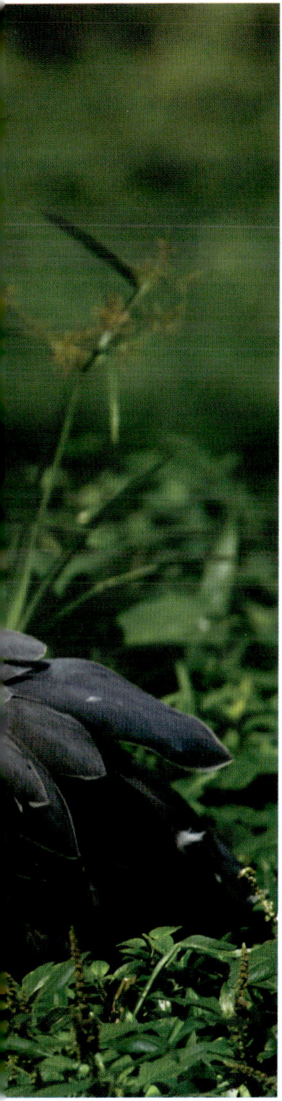

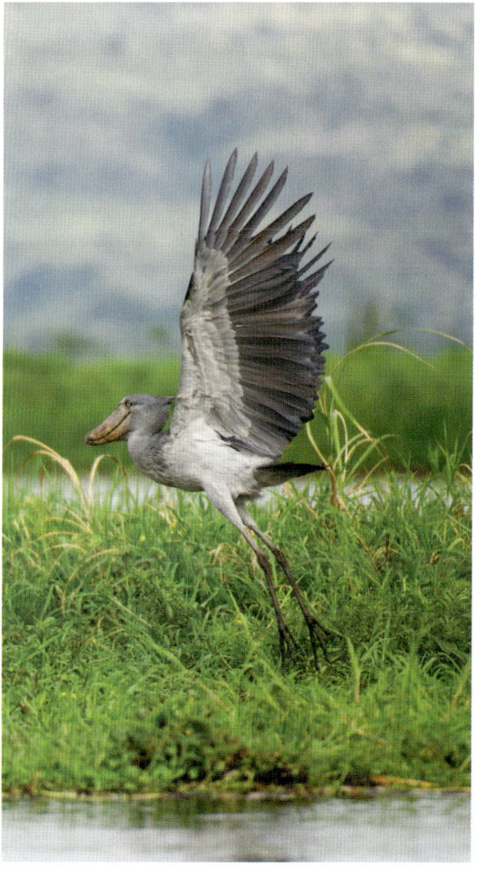

ALL PHOTOGRAPHS:

Shoebill

Named for its highly distinctive beak, this chunky bird lives in the wetlands around Africa's Great Lakes. The bill is adapted to scooping up fish in the muddy water. There are about 5,000 left in the wild, and numbers are dropping as available habitats are drained for agricultural use.

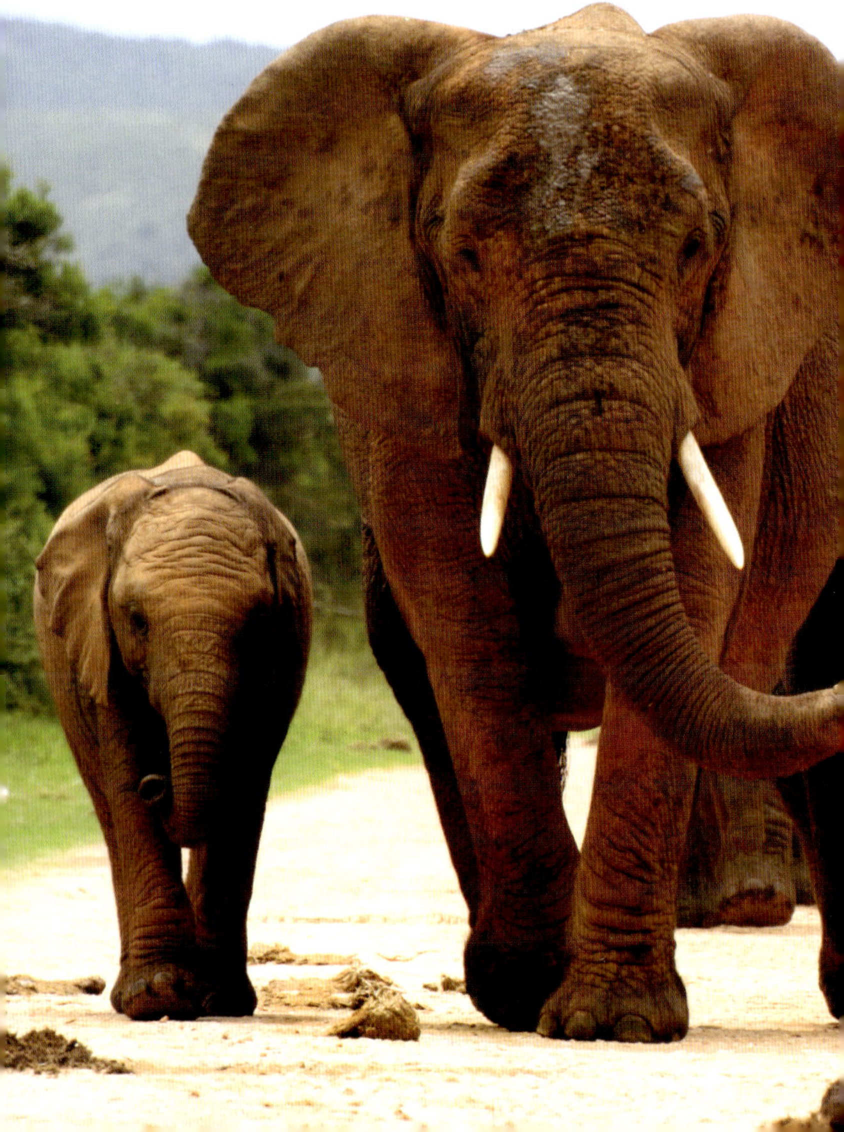

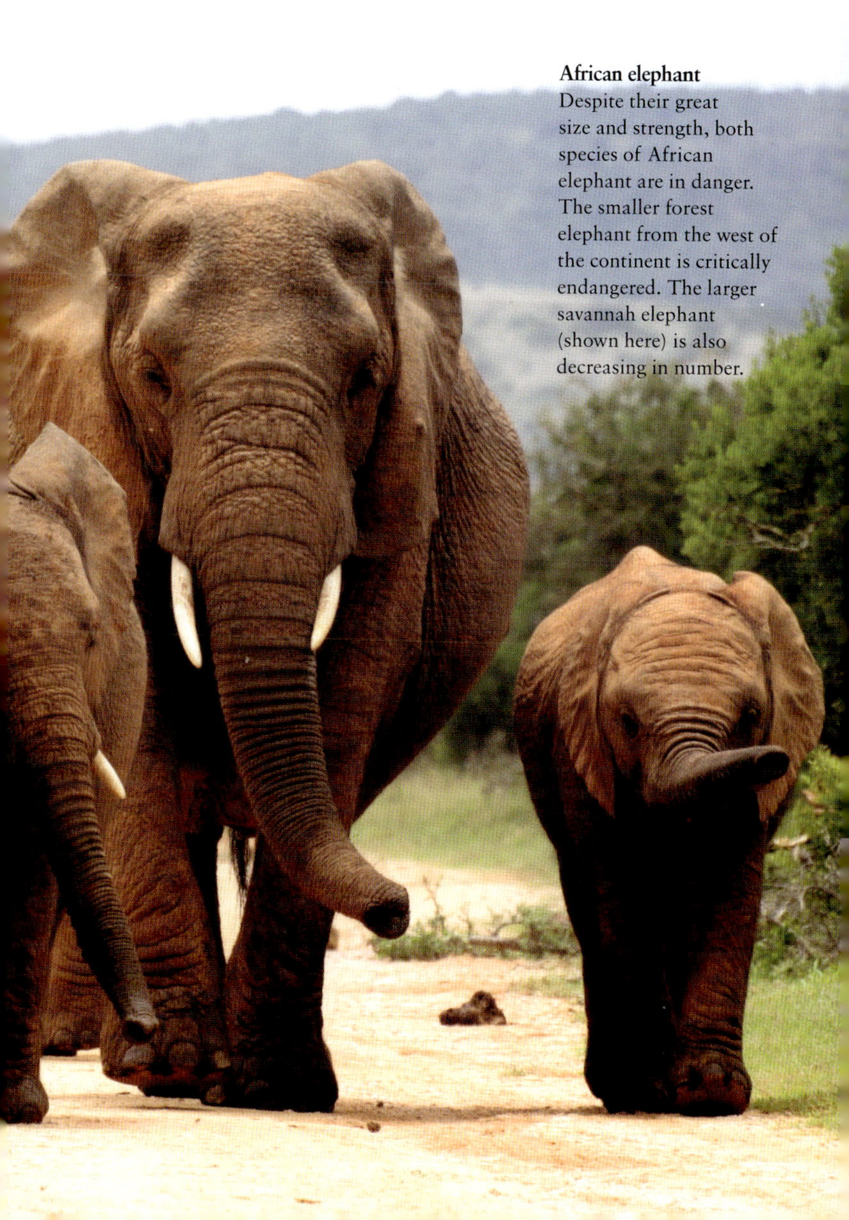

African elephant
Despite their great size and strength, both species of African elephant are in danger. The smaller forest elephant from the west of the continent is critically endangered. The larger savannah elephant (shown here) is also decreasing in number.

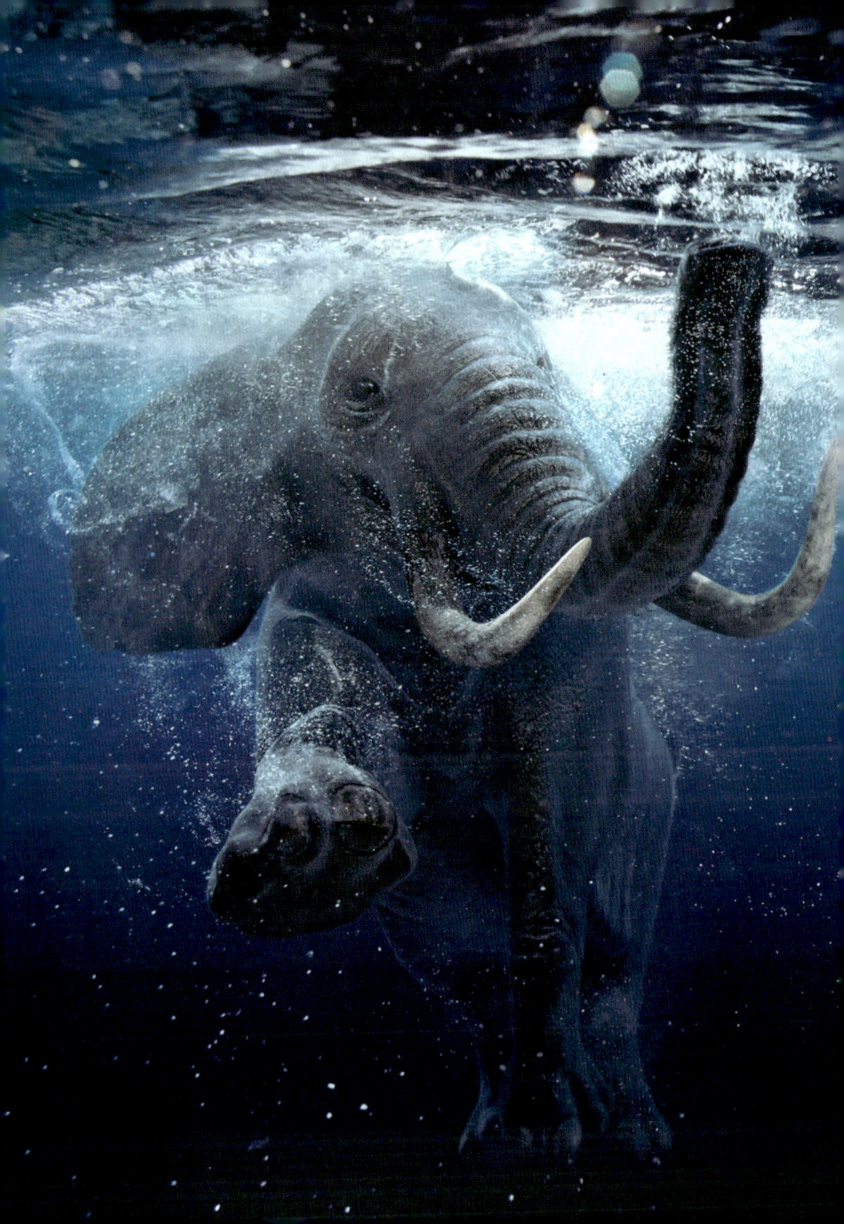

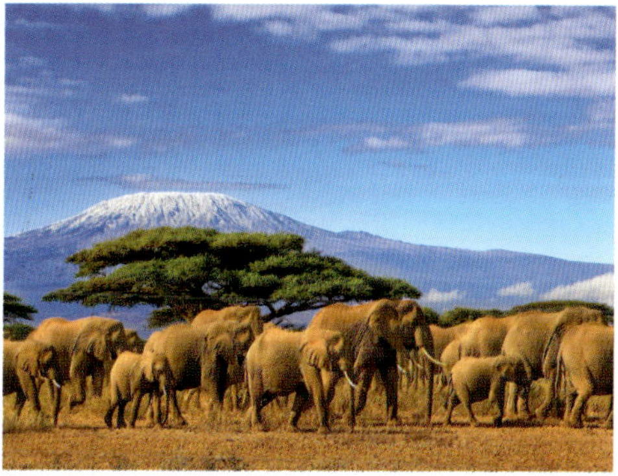

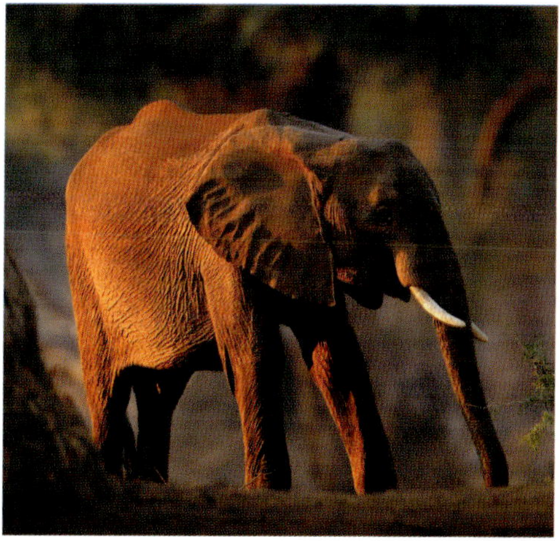

ALL PHOTOGRAPHS:

African elephant

The elephants are suffering from habitat loss, especially the forests of the Congo and West Africa. And wandering herds on the savannah are often in conflict with farmers as they feed on crops. However, the elephant's greatest threat is the poacher, who comes to kill the adults – or entire families – to steal their ivory tusks.

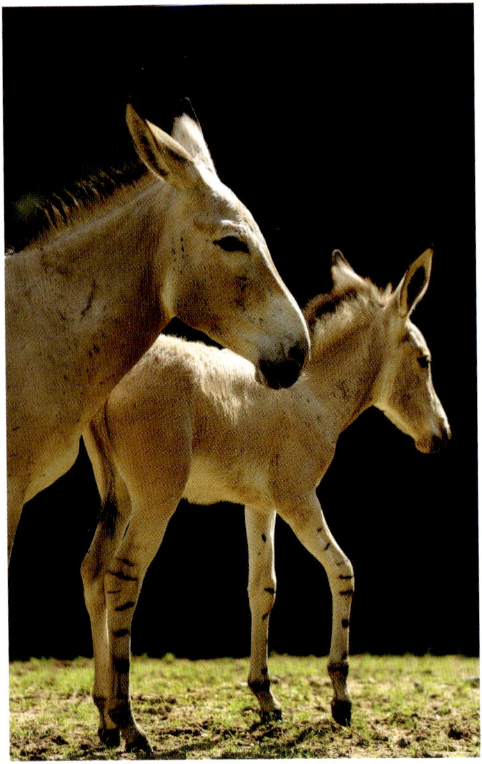

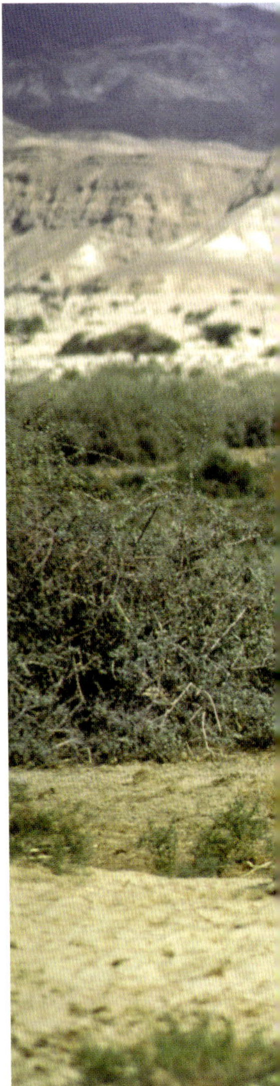

ALL PHOTOGRAPHS:

African wild ass

Surviving in a few patches along the western Red Sea coast, this wild ancestor of the donkey is critically endangered with fewer than 200 living outside of captivity.

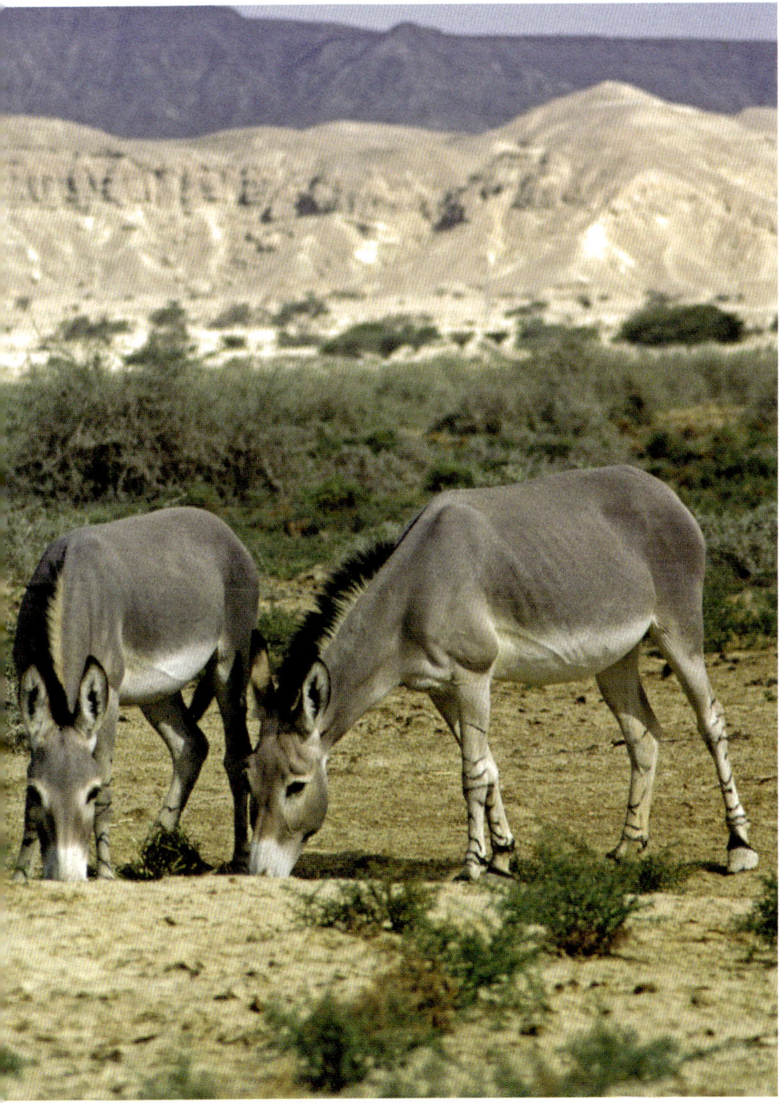

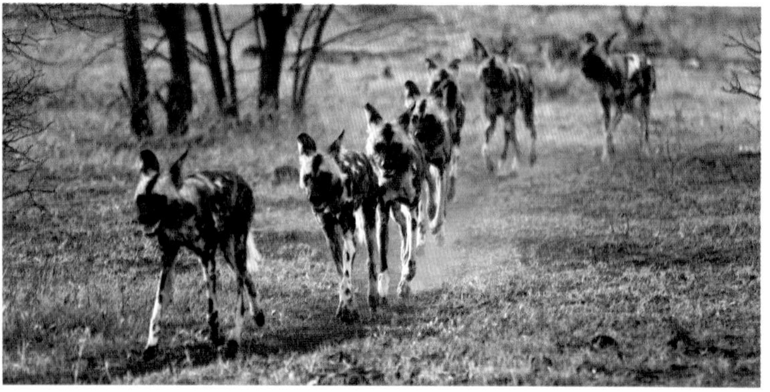

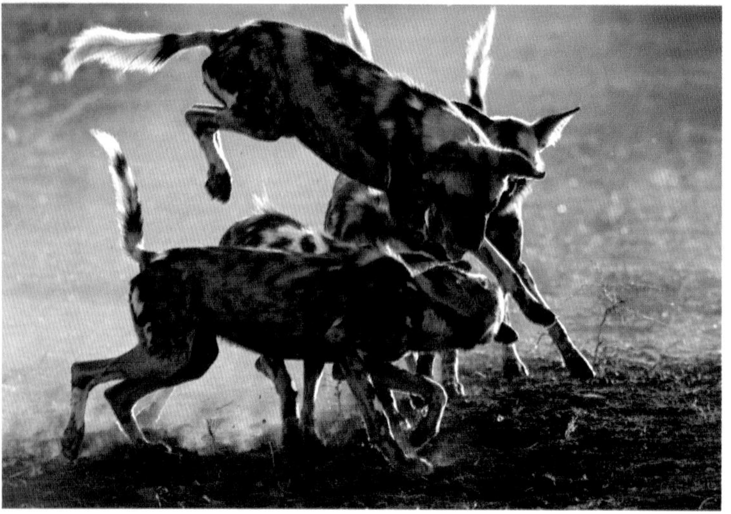

ALL PHOTOGRAPHS:
African wild dog
Found mostly in Botswana, with fragmented populations clinging on in other parts of the continent, this long-legged dog species is a pack hunter like the wolf, but is more closely related to the jackal. There are an estimated 1,400 left. The packs need a large territory to find enough food. They are being crowded out of the more fertile areas and now survive mostly in dry semi-desert regions.

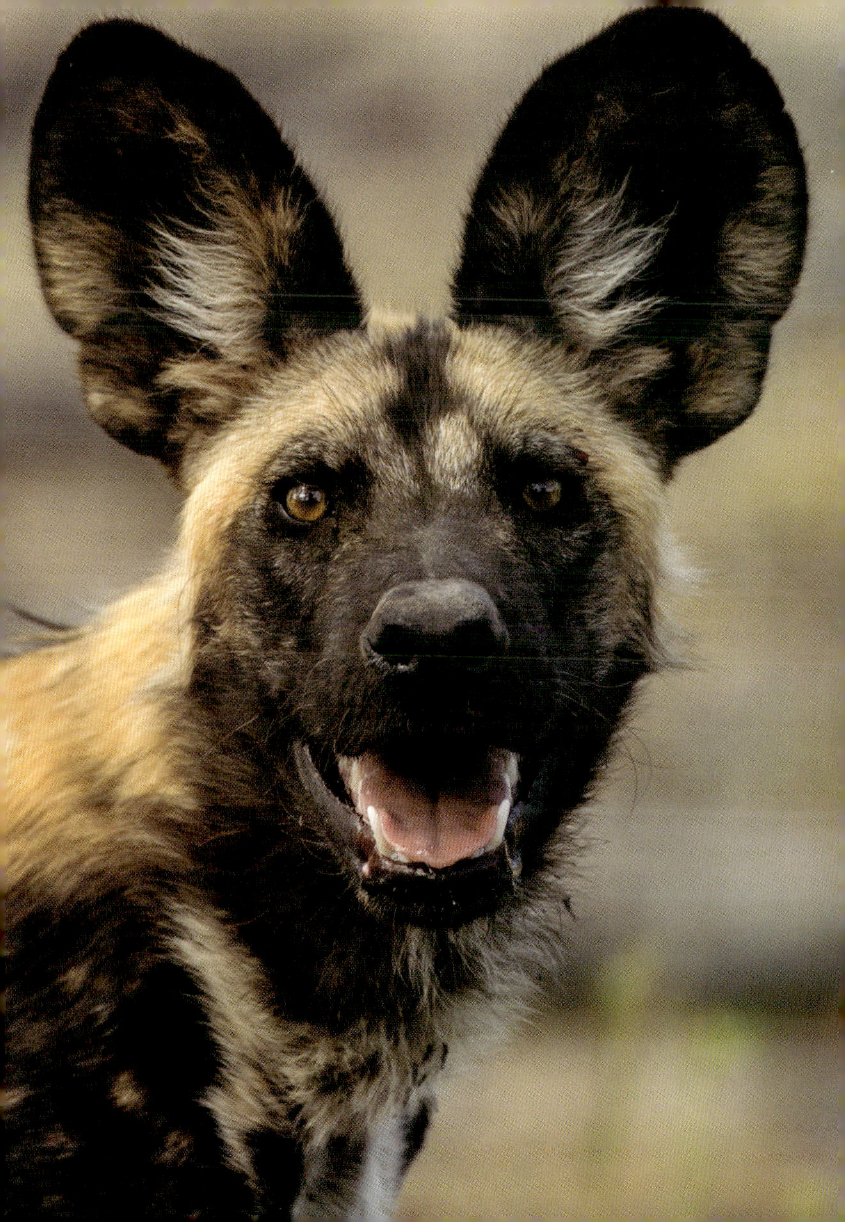

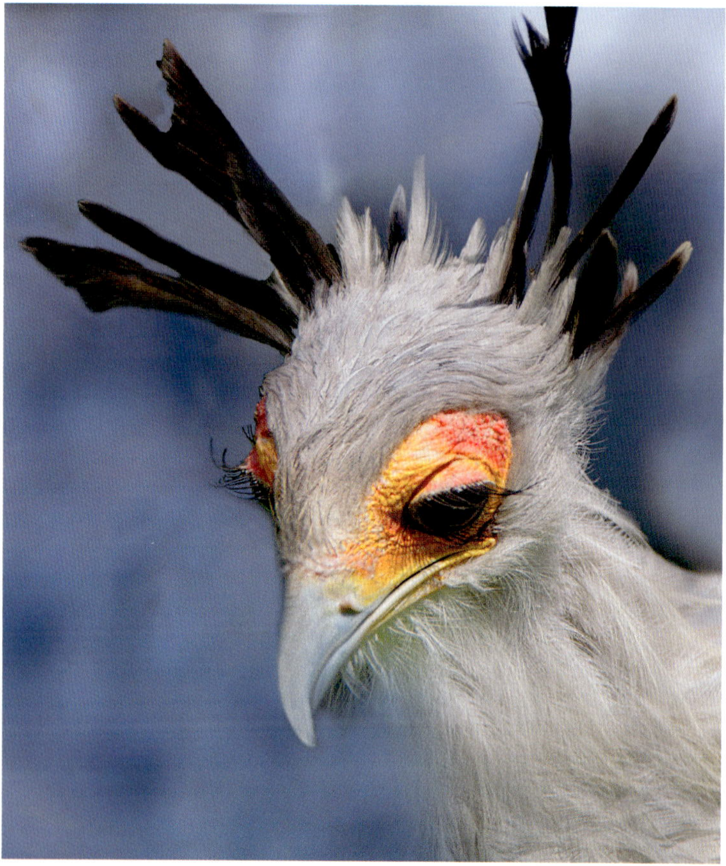

ALL PHOTOGRAPHS:

Secretarybird

Said by some to be named after the typewriter sound of its call, others suggest the way this insect-eating bird walks reminded people of the centaur archer Sagittarius, and that name was mangled over the years. The secretarybird has suffered a rapid decline in recent years as its habitat is fragmented by new roads.

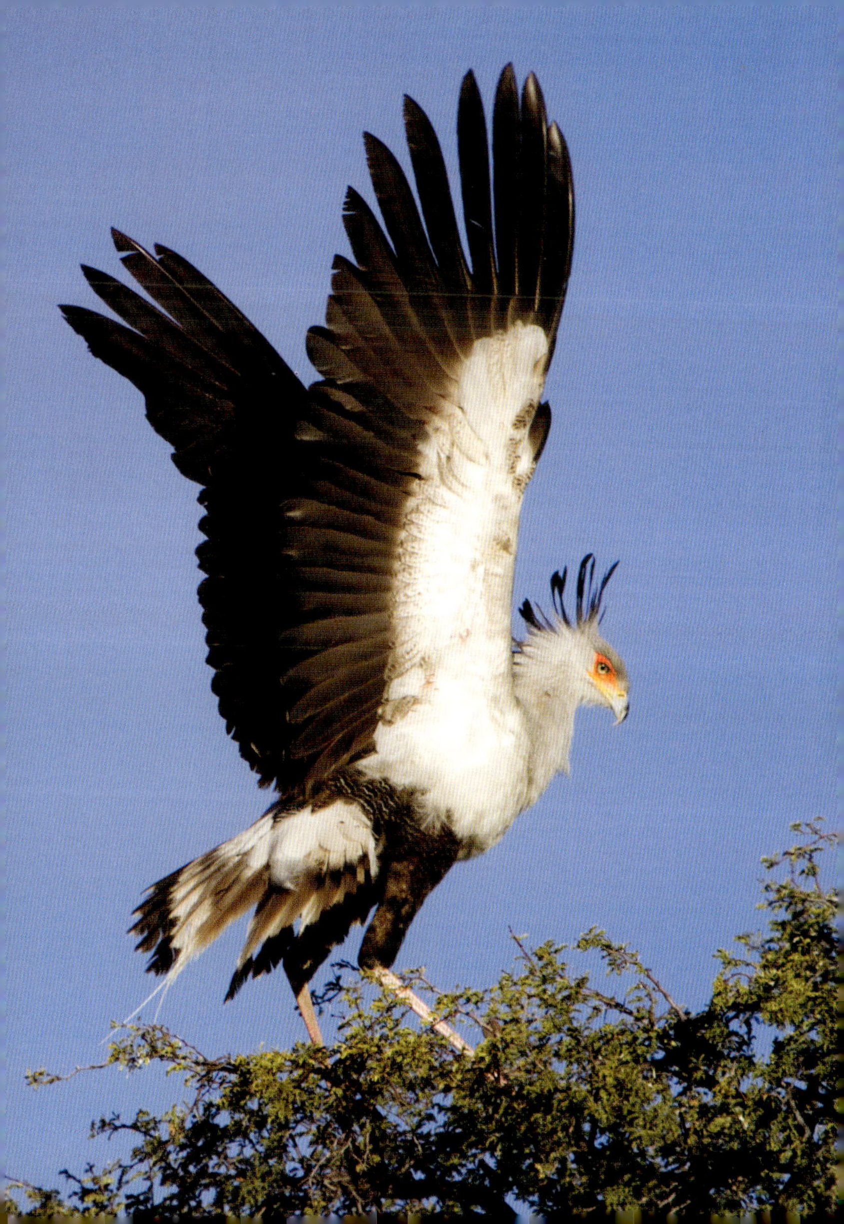

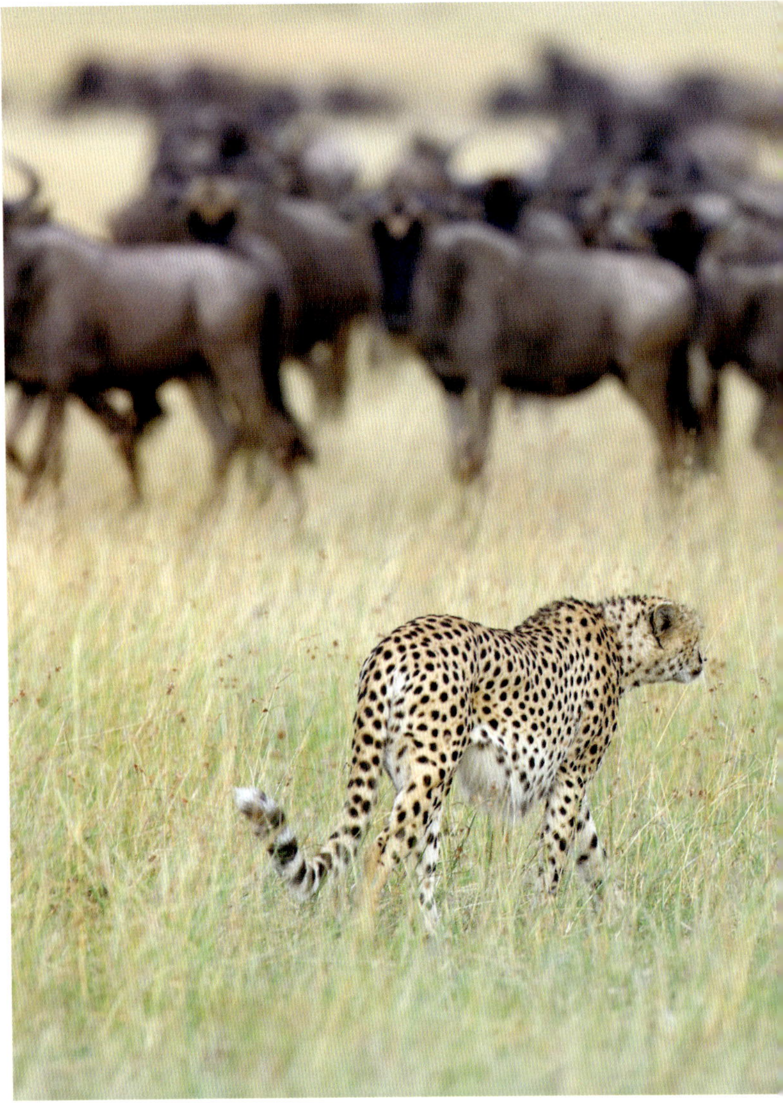

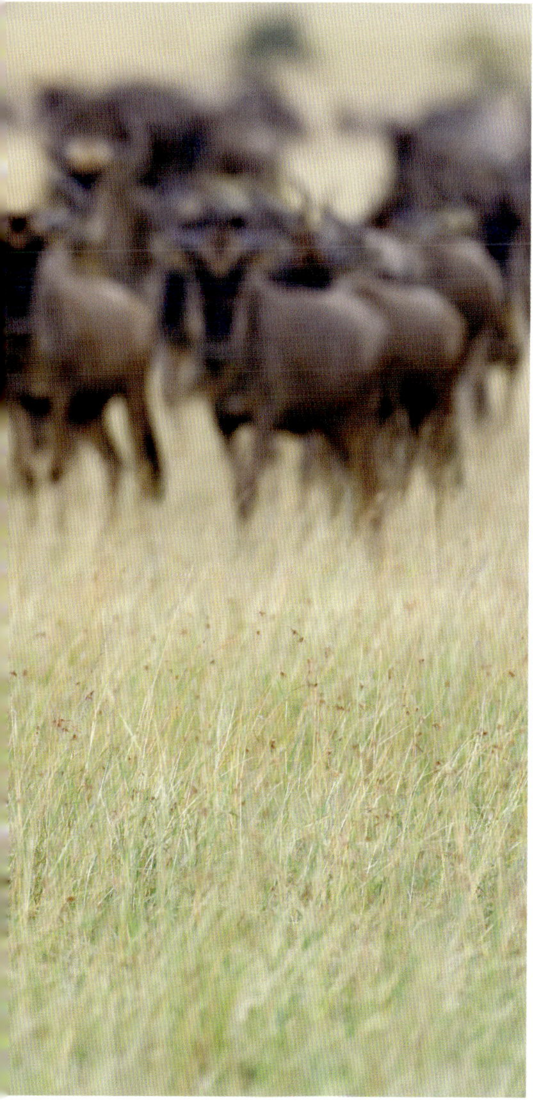

Cheetah
The fastest land animal of all, the graceful cheetah once lived across Africa and western Asia. It clings on in tiny numbers in Iran, and even in its stronghold of southern and eastern Africa there remain only 6,600 mature adults.

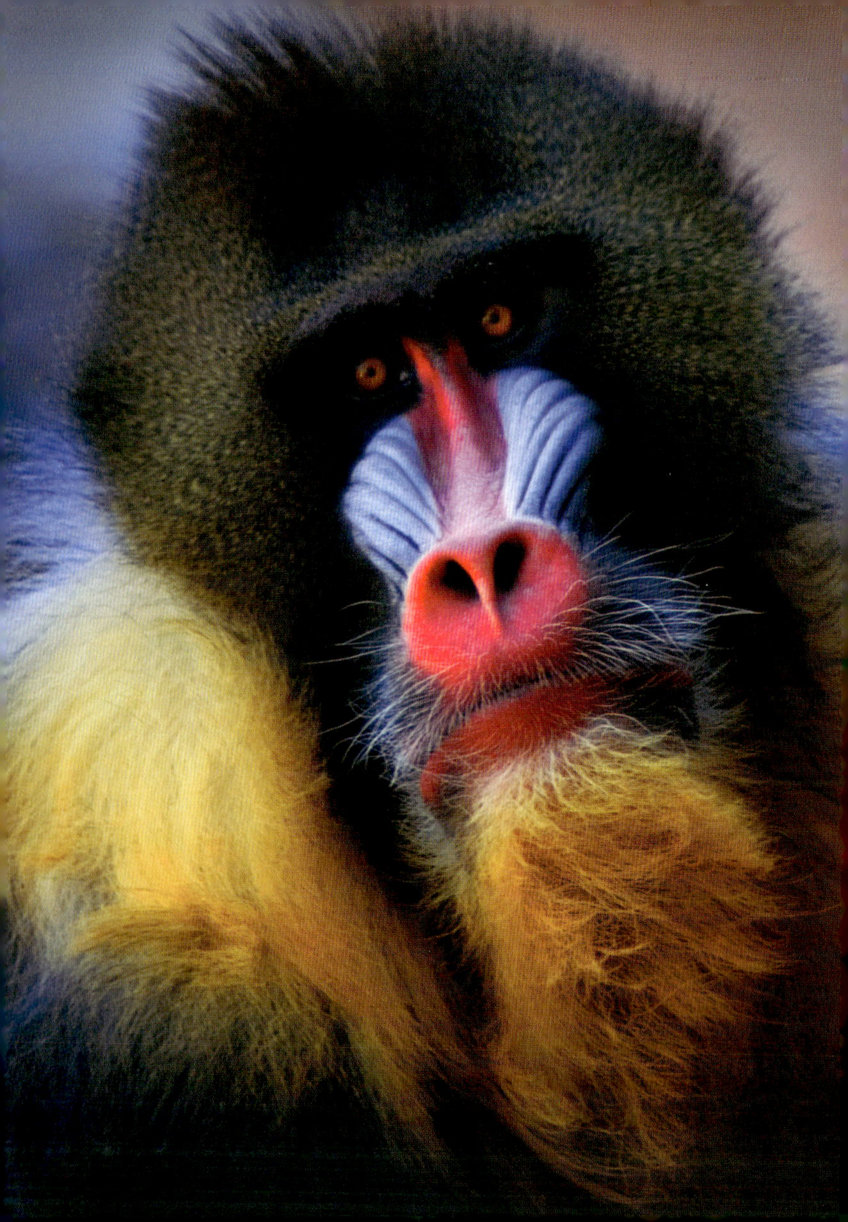

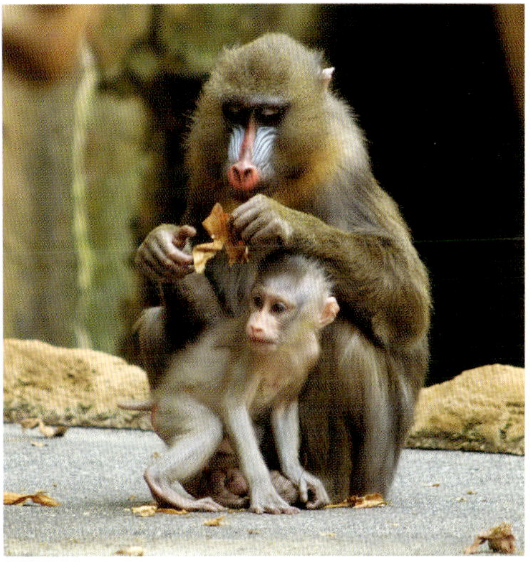

ALL PHOTOGRAPHS:
Mandrill
The largest monkey species of all, the mandrill lives in the mountain forests north of the mouth of the Congo. Their numbers in the wild are unknown since they live in the heart of dense jungles, but they are thought to be in decline. The male mandrills are famed for having impressive facial marking that flare with colour in times of tension and excitement.

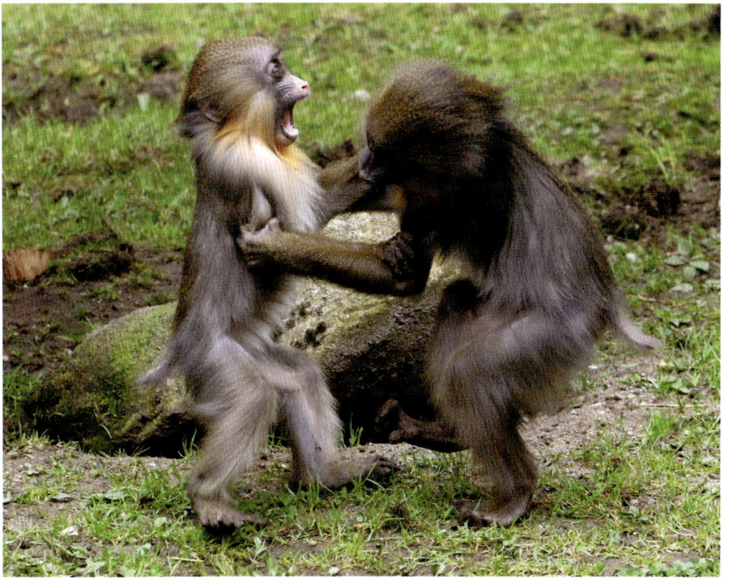

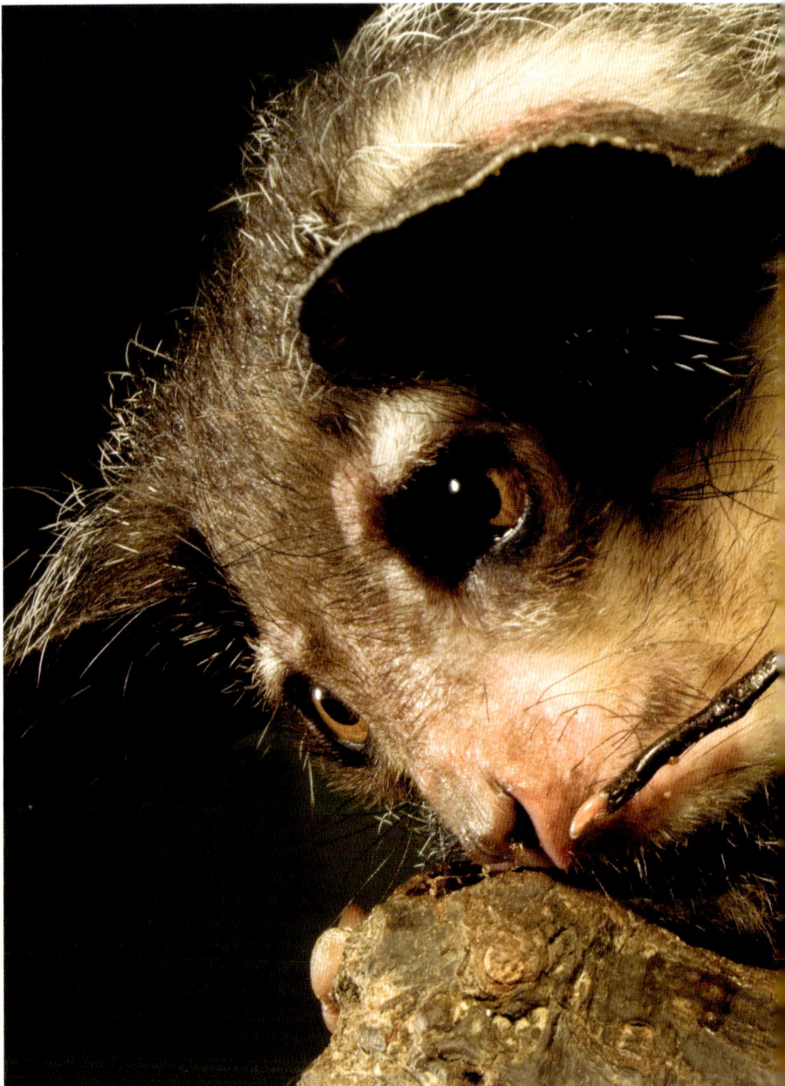

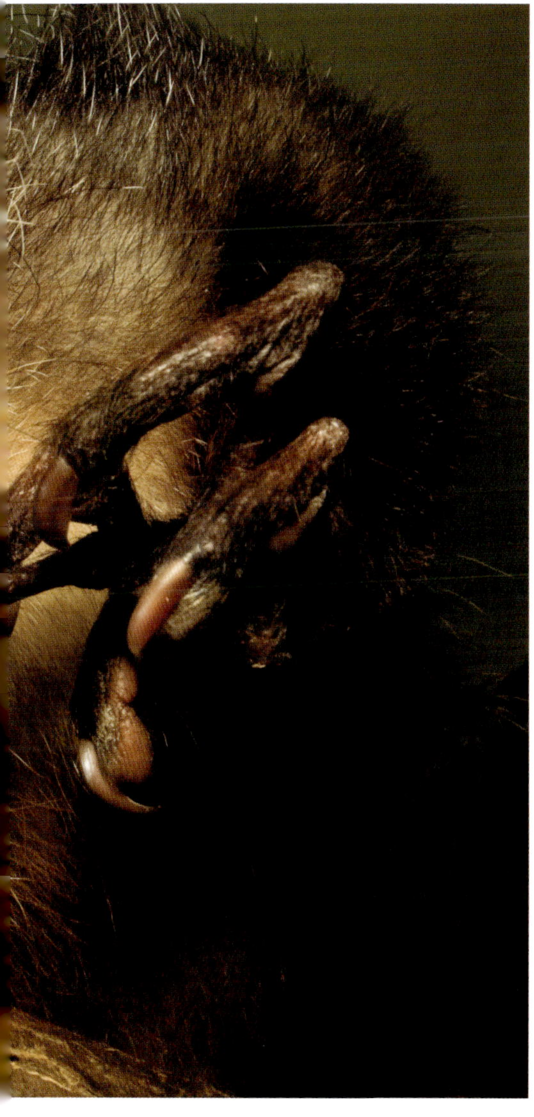

LEFT:

Aye-aye

All the lemurs in Madagascar are endangered in some way, but this highly unusual nocturnal species is especially vulnerable. It uses its elongated finger to tap on tree branches listening out for hollow sounds that show a juicy grub is lurking under the bark. The animals' forest home is under threat, and local people see the ghoulish creature as evil and must kill it on sight.

OVERLEAF:

Black rhinoceros

Hunted relentlessly for its nose horn, conservation efforts saved the black rhino from extinction. There are now 3,500 black rhinos in the wild with numbers rising. The black species is contrasted with the white rhino, which is not white at all but instead has a wide mouth for cropping grass. The black rhino's more beak-like mouth is suited for browsing leaves from woodland shrubs.

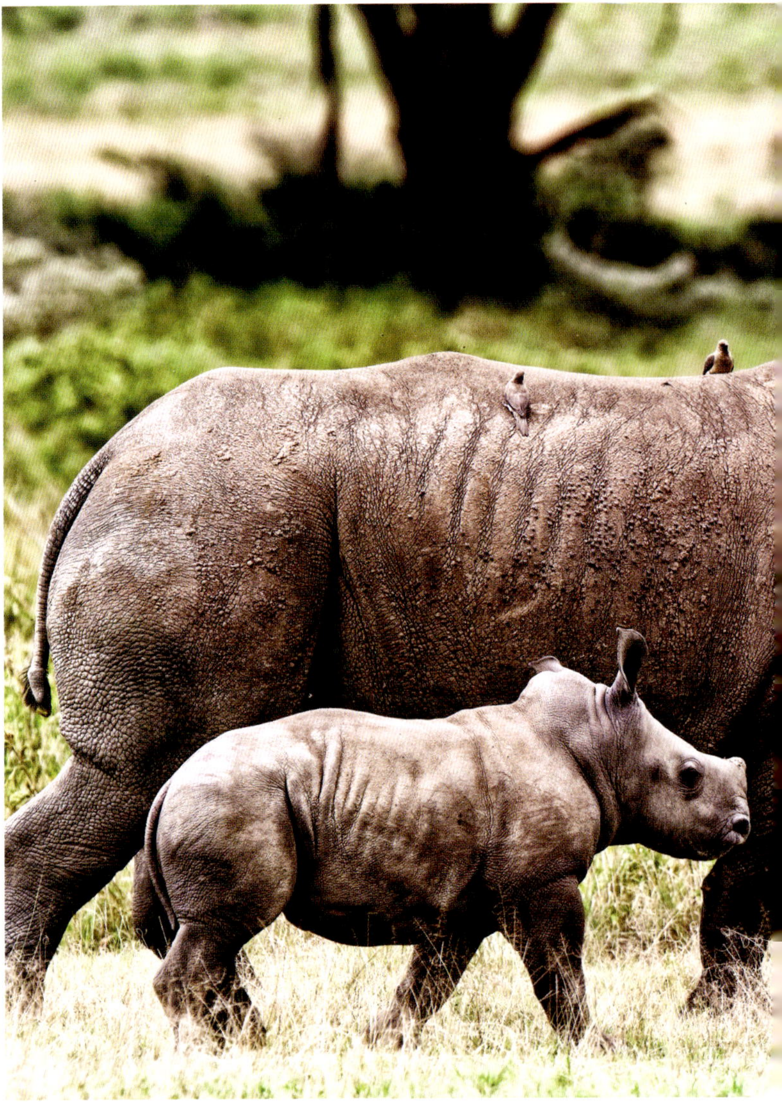

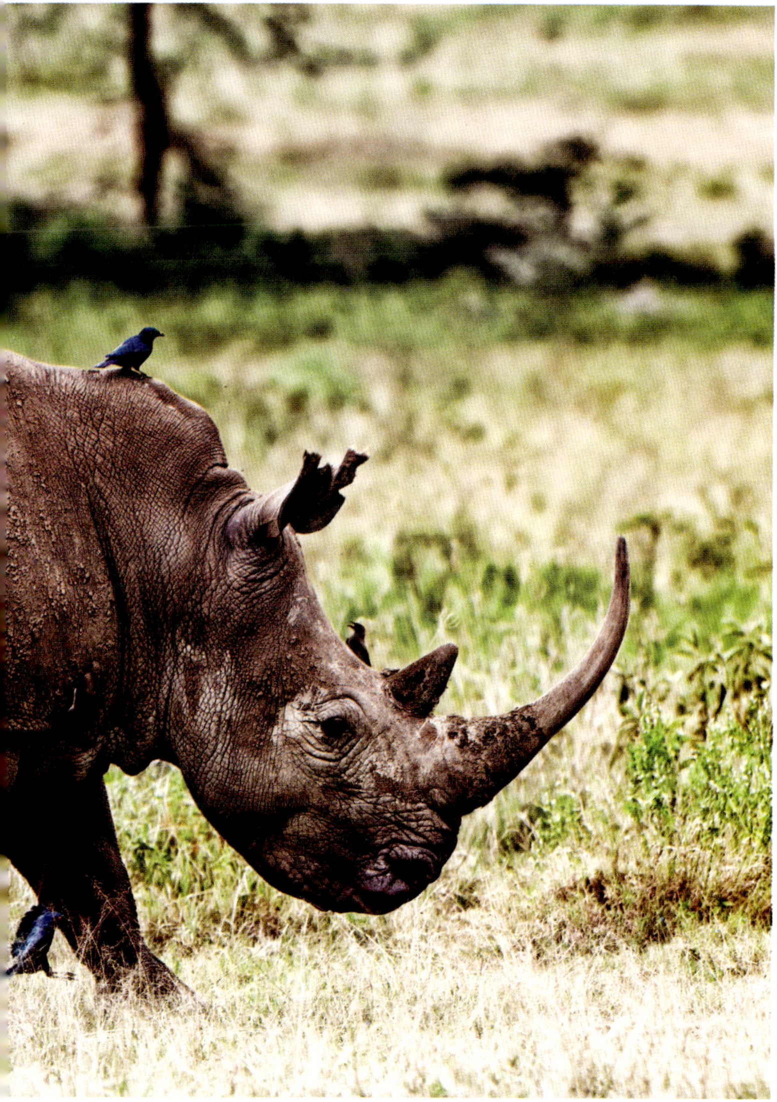

ALL PHOTOGRAPHS:

Rodrigues flying fox

Found only on a remote Indian Ocean island, this is one of the world's largest bats, with a wingspan of 90cm (35in). Being so isolated, the bat was very vulnerable to habitat loss on the island. By the 1990s, the population hit a nadir of 1,000, but has since recovered to around 20,000.

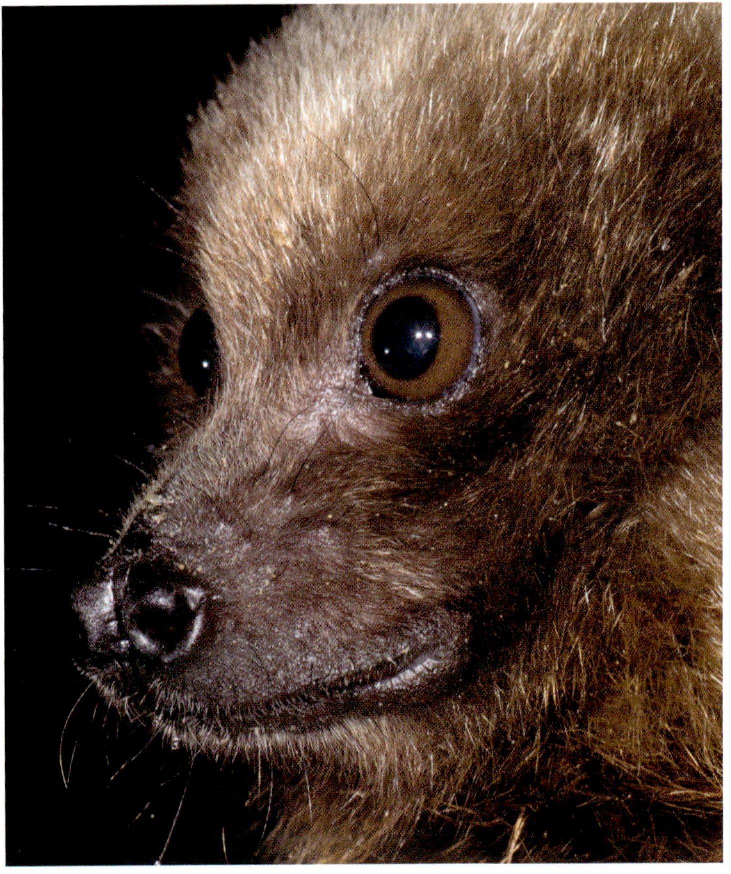

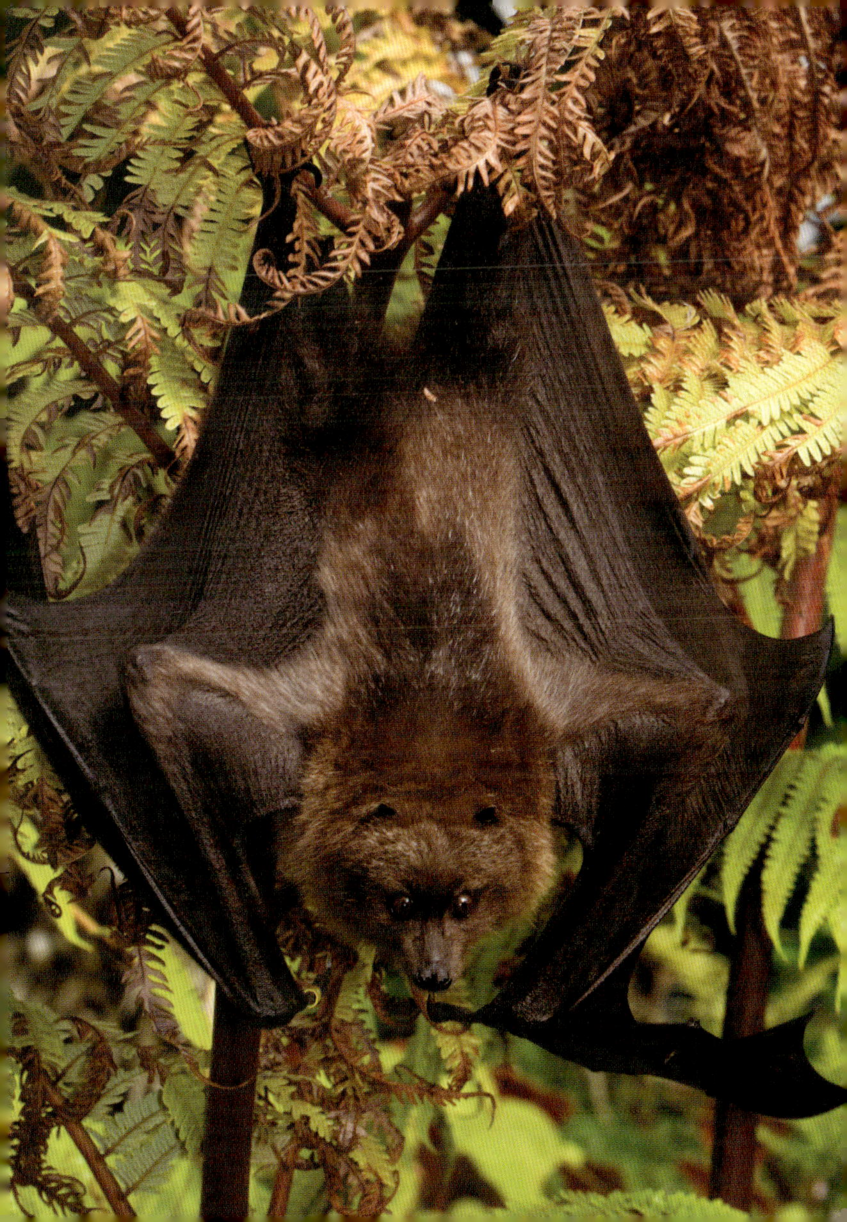

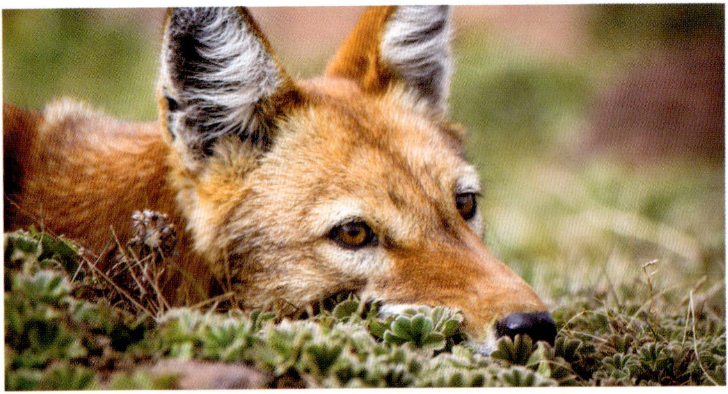

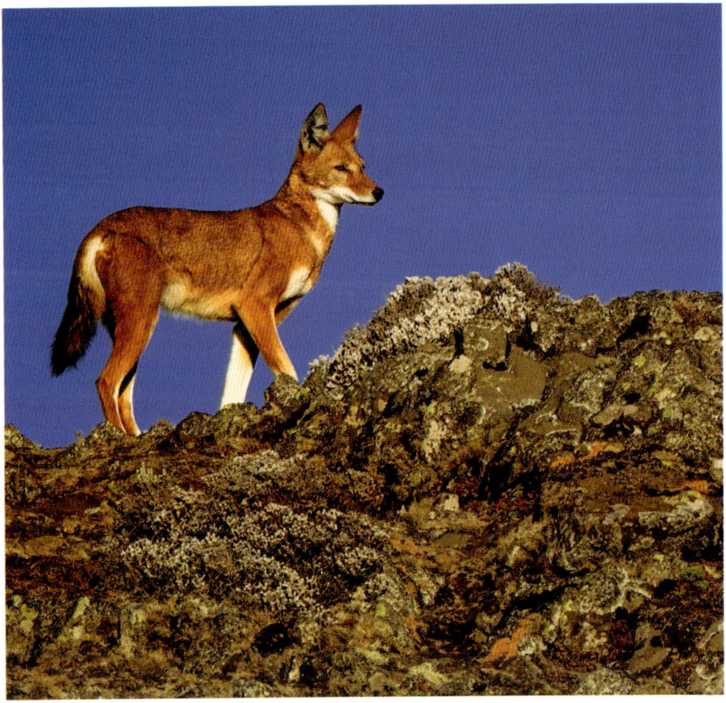

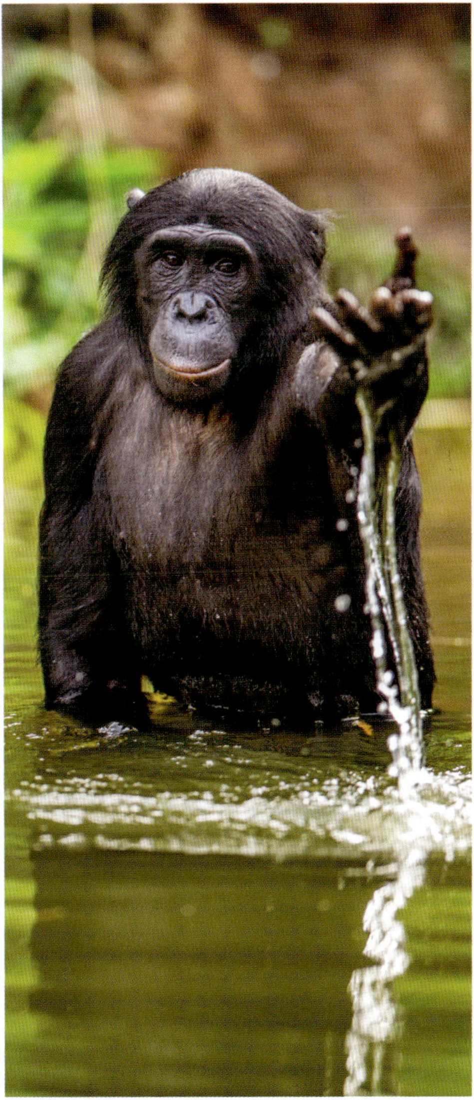

OPPOSITE TOP AND BOTTOM:

Ethiopian wolf

Despite the name, this dog species more closely resembles the jackal. It lives in a small pack but always hunts alone in the Ethiopian Highlands, mostly preying on burrowing mole rats. The species is at risk from interbreeding with feral dogs that arrive from the human settlements in the valleys below.

LEFT:

Bonobo

Also called the pygmy chimpanzee, this African ape is our closest relative. Found only in the forests of western Congo, the wild population is estimated to be under 20,000.

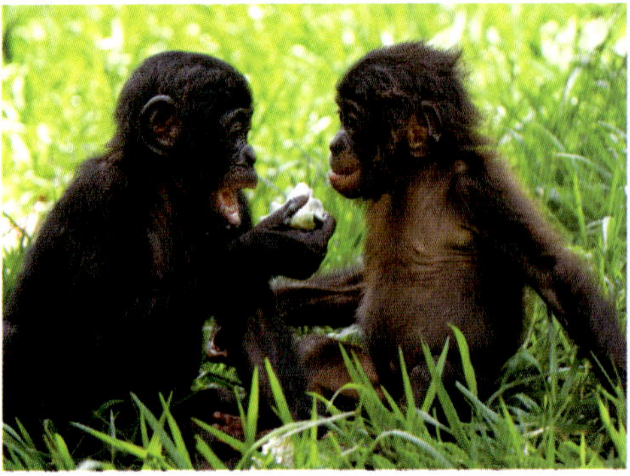

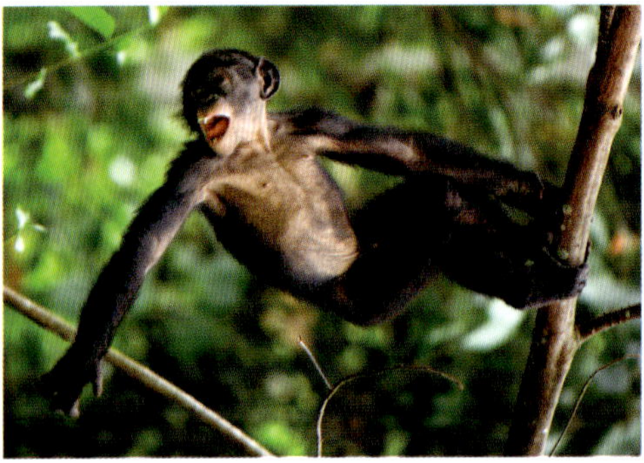

ABOVE TOP AND BOTTOM:

Bonobo

It is not easy to count the number of bonobos in the dense jungle. Groups are frequently on the move, so instead conservationists count their nests, made fresh each night on the ground or low branches and then abandoned in the morning.

RIGHT:

Common chimpanzee

Standing head and shoulders above the bonobo, the common chimpanzee lives in another part of the Congo Basin, with other populations over in West Africa. Poaching and habitat loss mean the 300,000-strong wild population is going down.

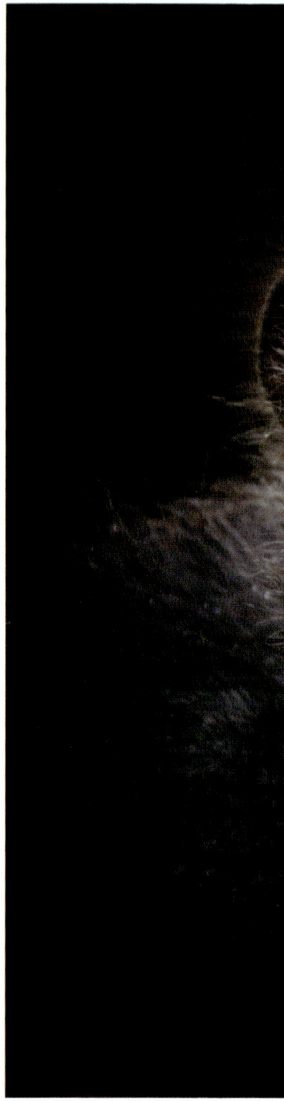

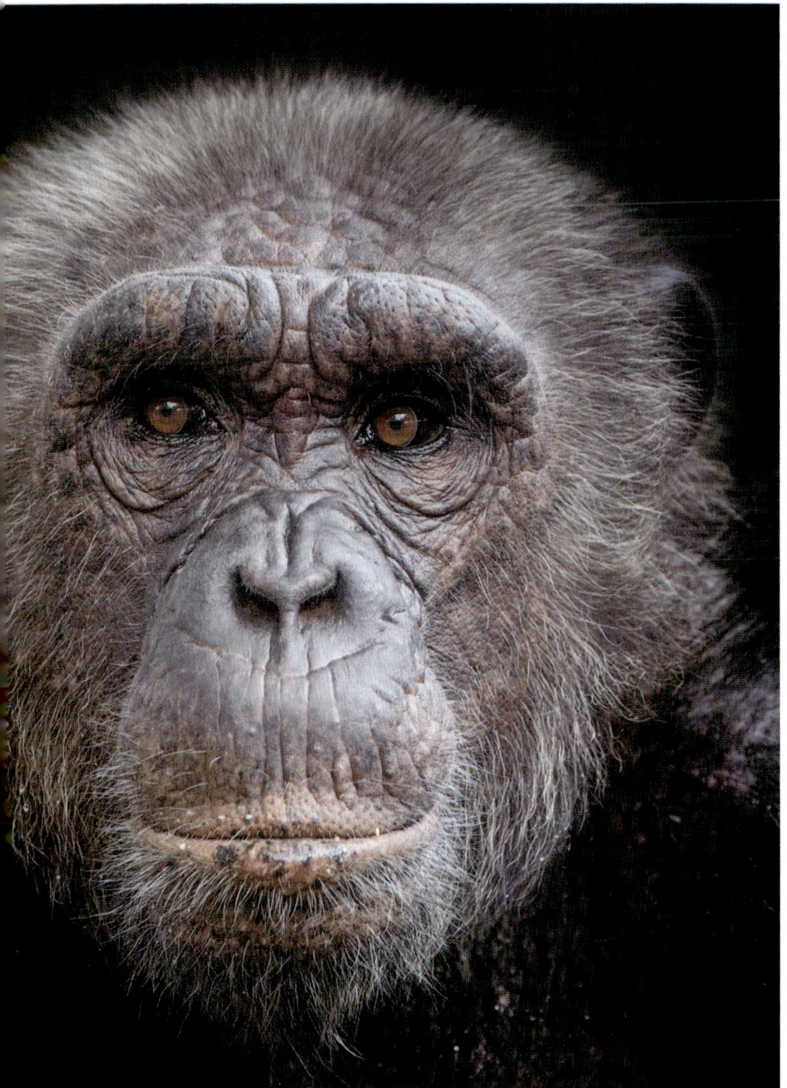

Chimpanzee

In the 1930s there were one million chimpanzees in Africa. Their decline is due mostly to the reduction in suitable rainforest where the apes feed on leaves, fruits, insects and occasionally flesh of prey killed by a hunting party. Contact with humans has also spread diseases among the apes, and they are still hunted for meat or captured as pets.

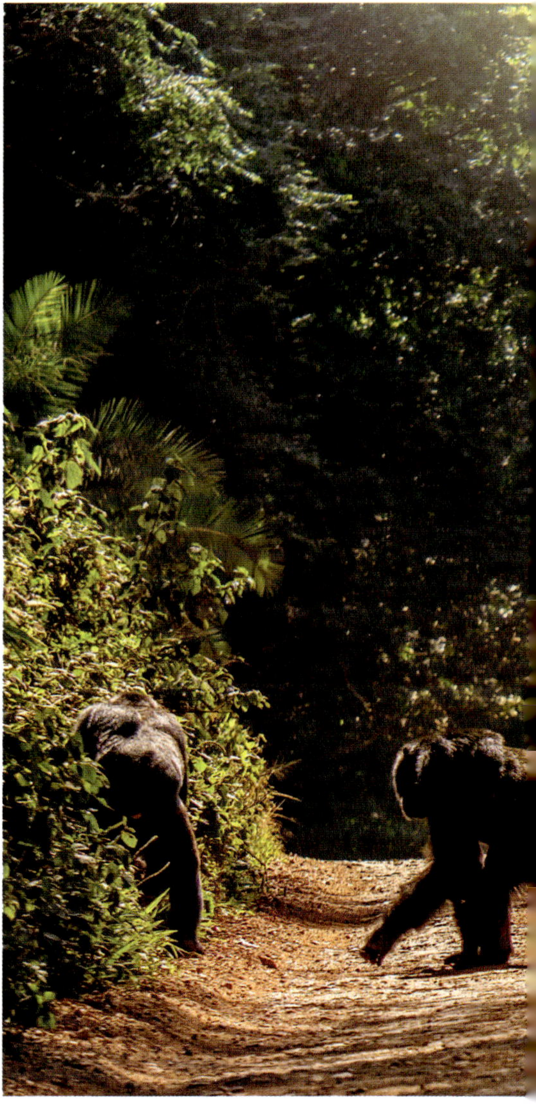

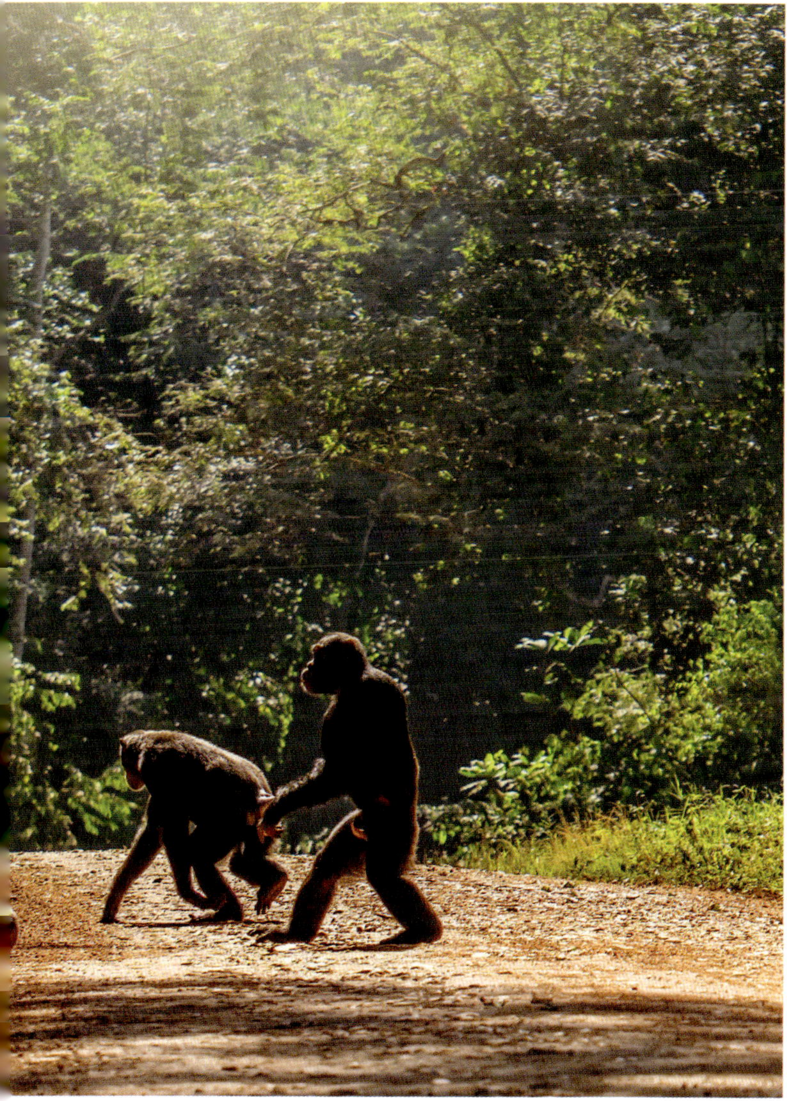

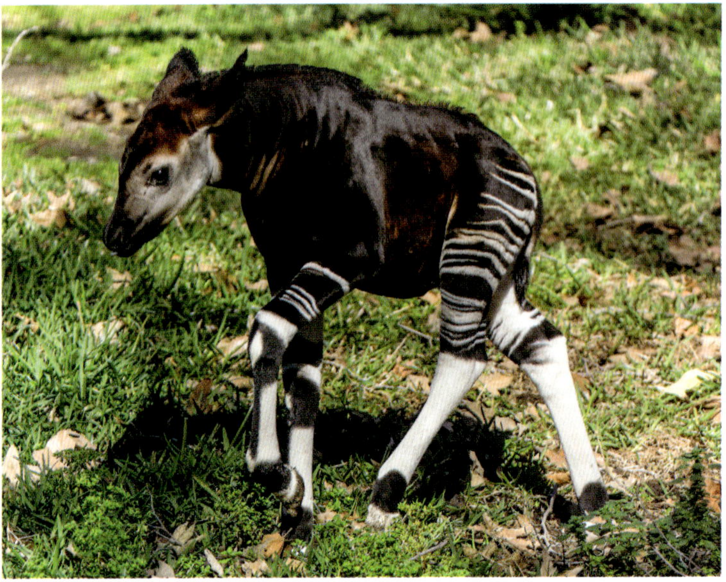

ALL PHOTOGRAPHS:

Okapi

The closest relative of the giraffe, the okapi is found in decreasing numbers in the eastern part of the Congo Basin. Estimates of wild numbers range from 10,000 to 50,000. The striped rump helps the large animal blend into the undergrowth, as it strips leaves from trees with its long tongue.

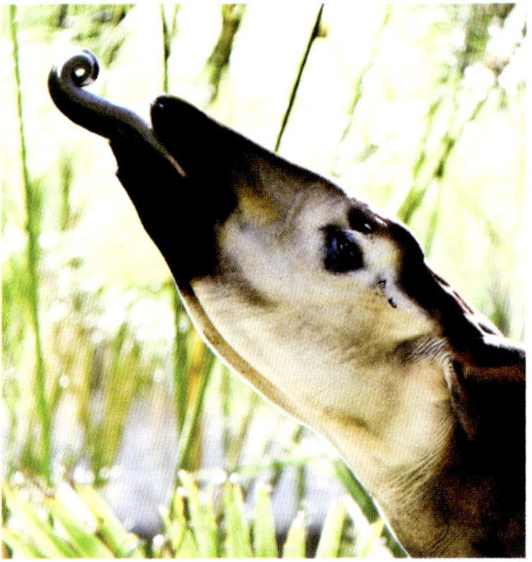

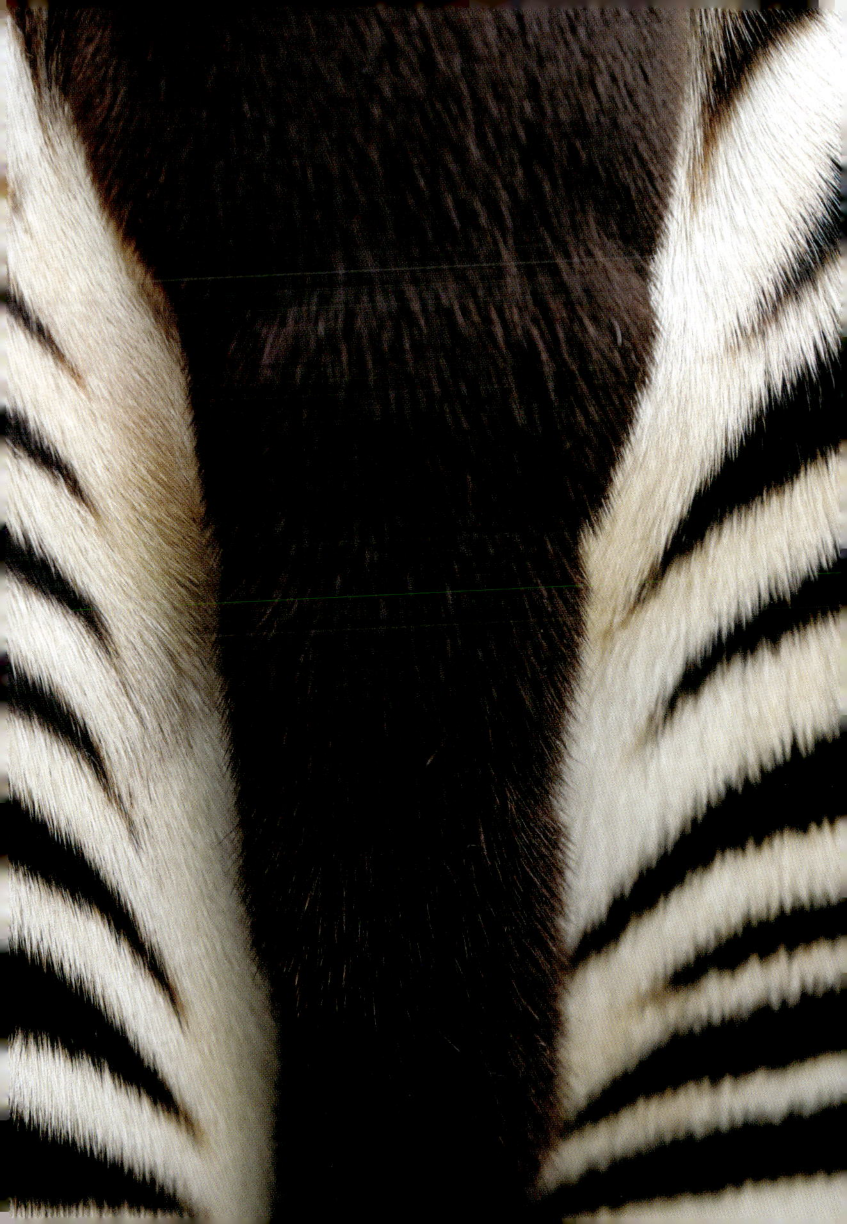

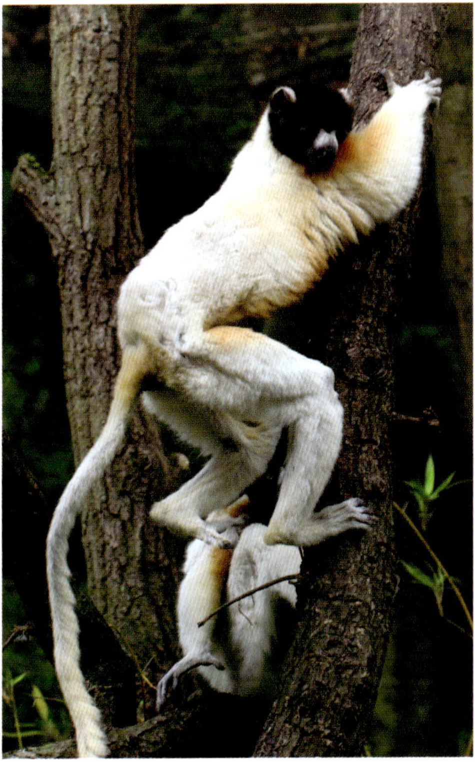

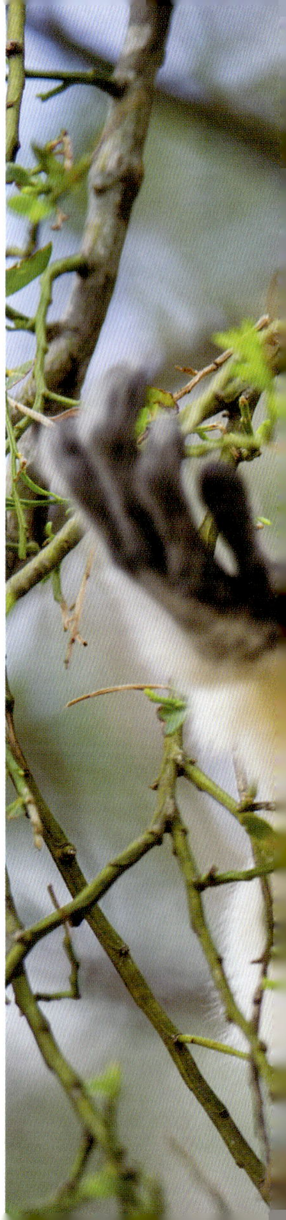

ALL PHOTOGRAPHS:

Crowned sifaka

Pronounced 'shi-far-kuh', these large lemurs are famed for the way they leap between the tall trunks of their forest home in Madagascar. As these forests thin and fragment, the sifakas are put under pressure. Unlike other lemurs, these species – the black crowned and golden crowned, respectively – are unable to walk on the ground, just hop, so cannot move across gaps in the forest.

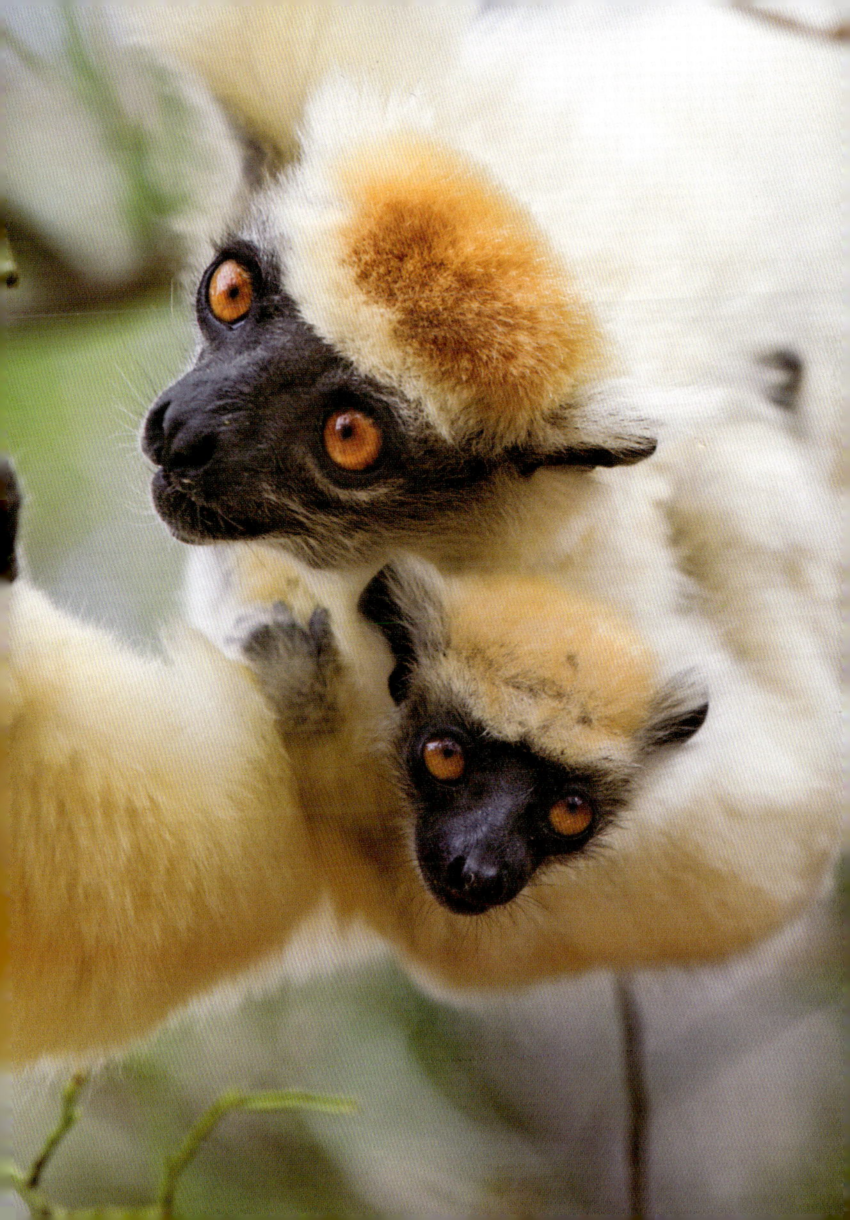

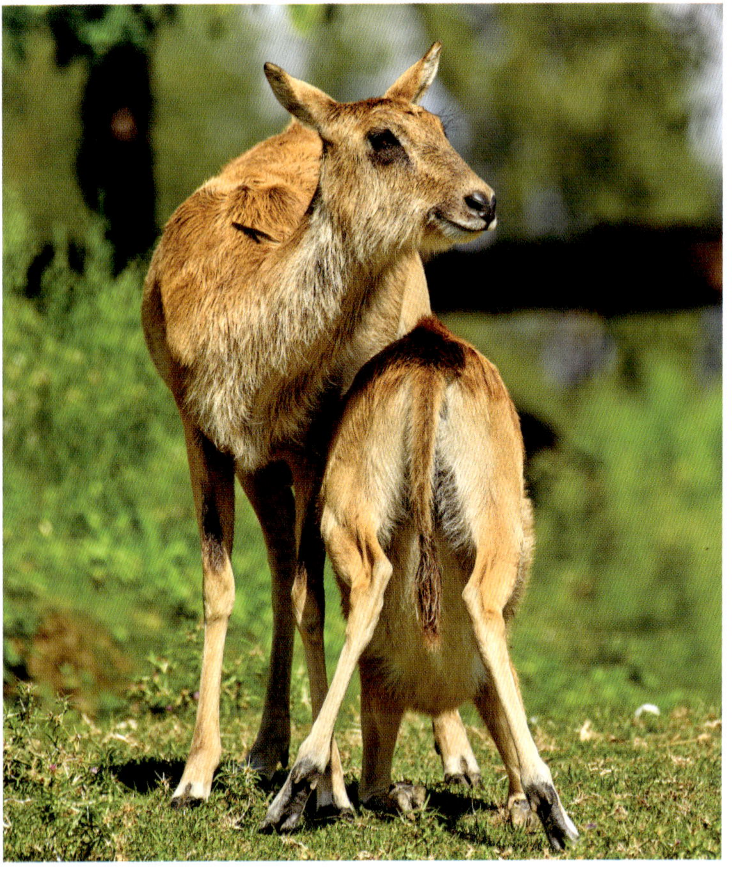

ABOVE:

Nile lechwe

The brown coats show these are female lechwes, with the males having a black coat. These antelopes live in the swamps of South Sudan formed by the White Nile. This habitat is shrinking and so is the lechwe population.

OPPOSITE:

Mountain gorilla

A subspecies of the critically endangered eastern lowland gorilla, this small band of herbivorous primates lives in the cloud forests that straddle the borders of Uganda, Congo and Rwanda.

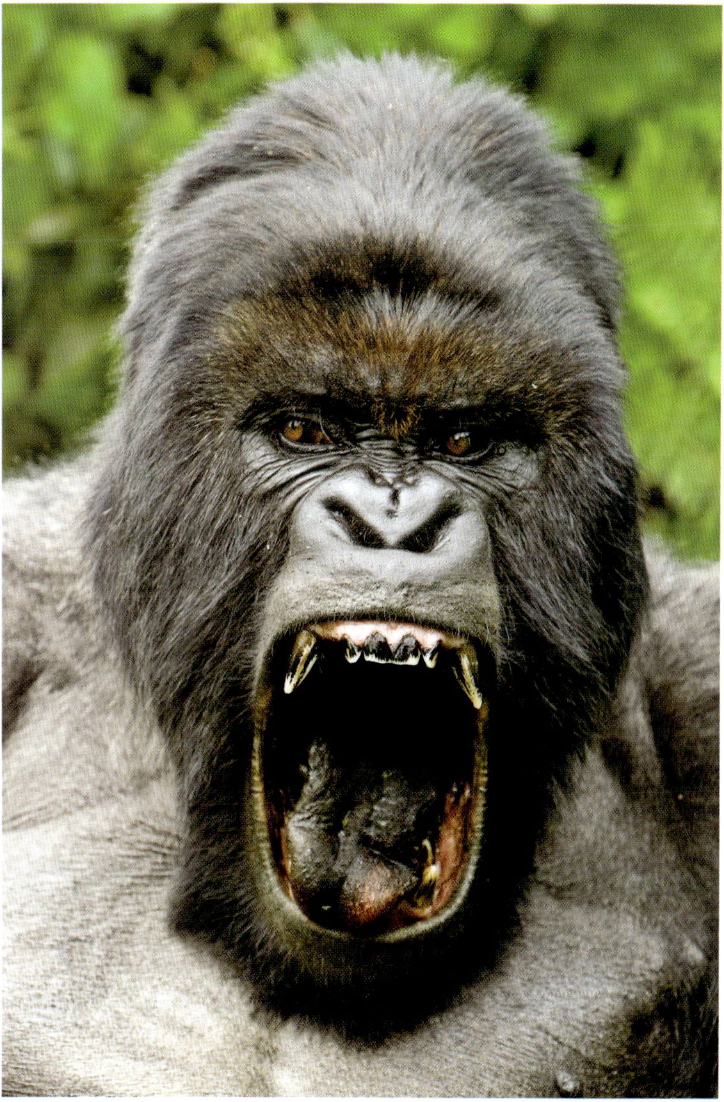

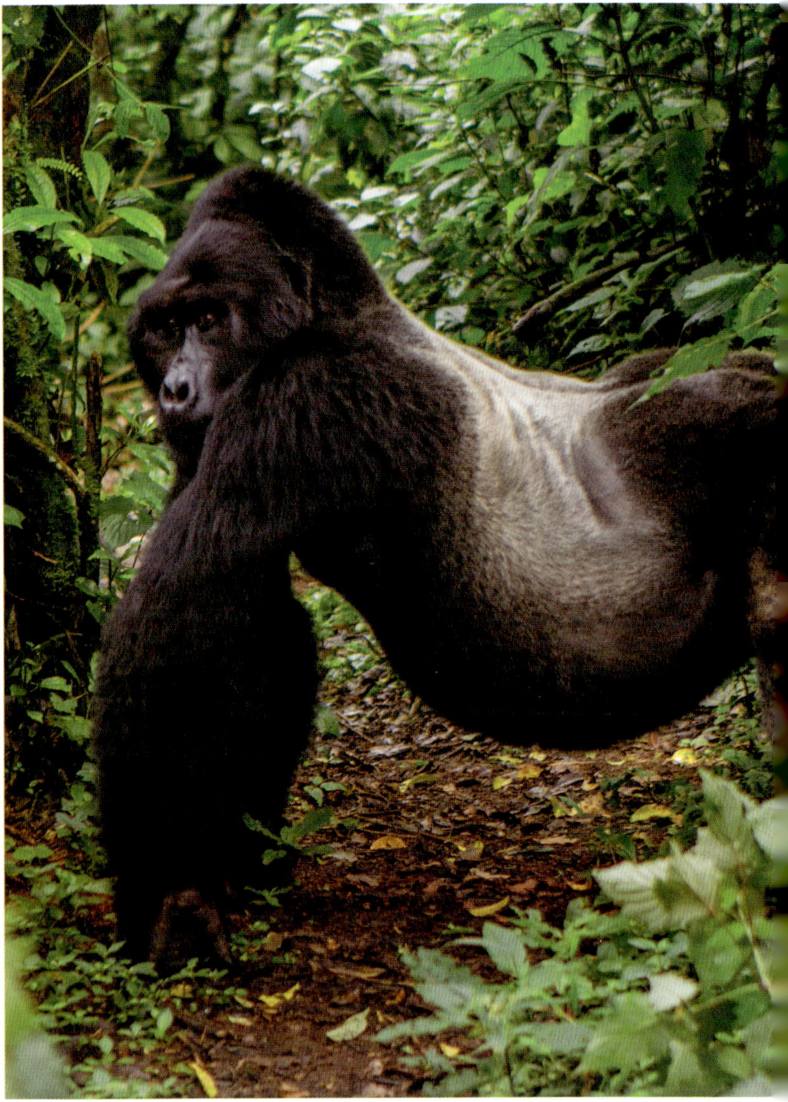

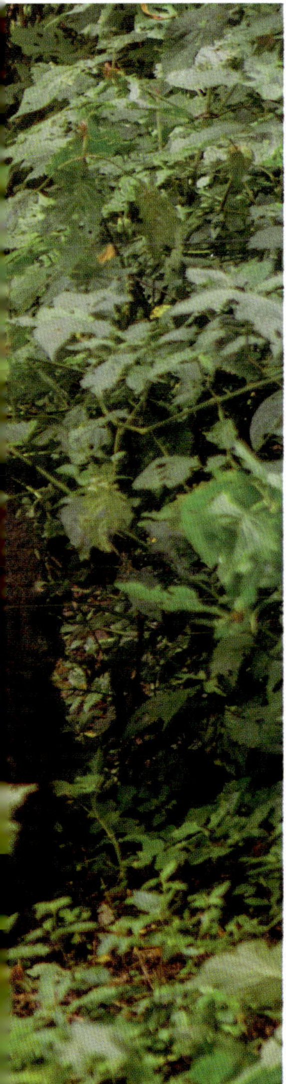

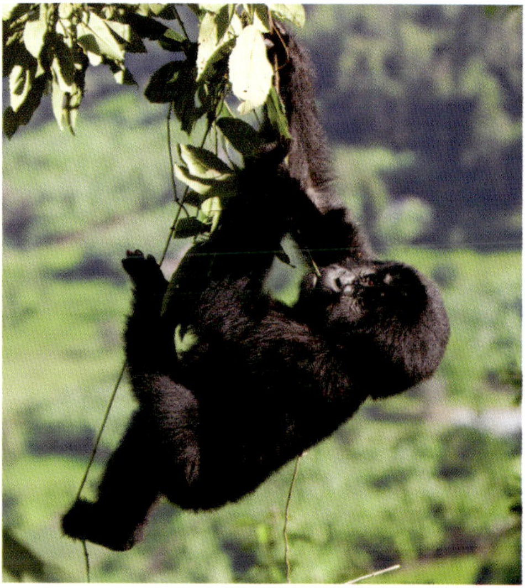

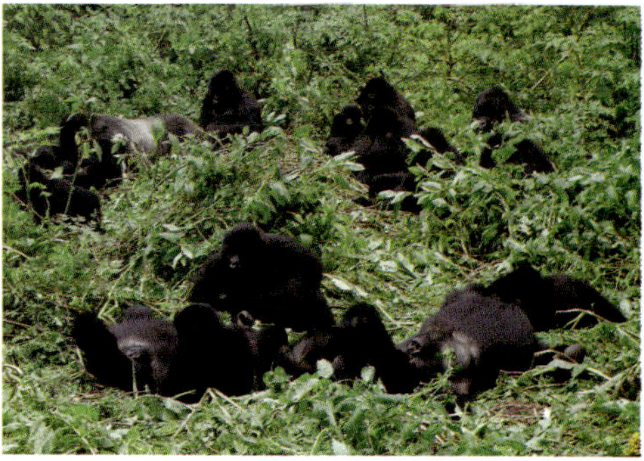

ALL PHOTOGRAPHS:

Mountain gorilla

Under attack from guerilla armies, poachers and illegal loggers, the mountain gorilla population fell to 250 in the early 1980s, but has since recovered to 1,000 and is continuing to rise as its habitat is now protected as a valuable ecotourism resource.

ALL PHOTOGRAPHS:

Pygmy hippopotamus

At first glance, there is little to show how these amphibious mammals differ from their more common cousins found across African waterways. However, this species, which lives in only a handful of numbers along the West African coast, is only half as tall as its giant cousin and is barely a quarter of its weight. Loss of the coastal forest means there are just 2,500 adults in the wild, at the most.

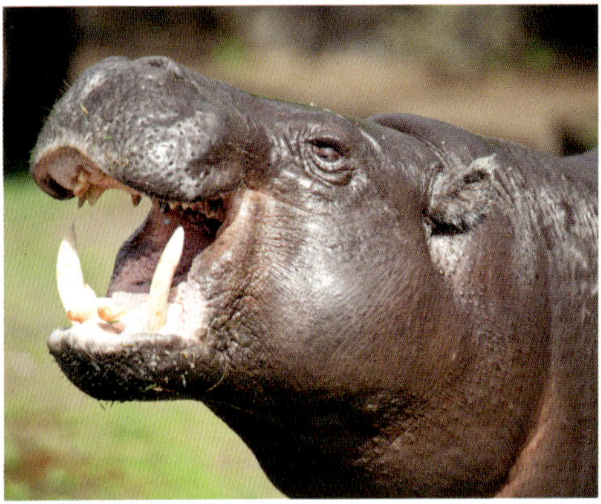

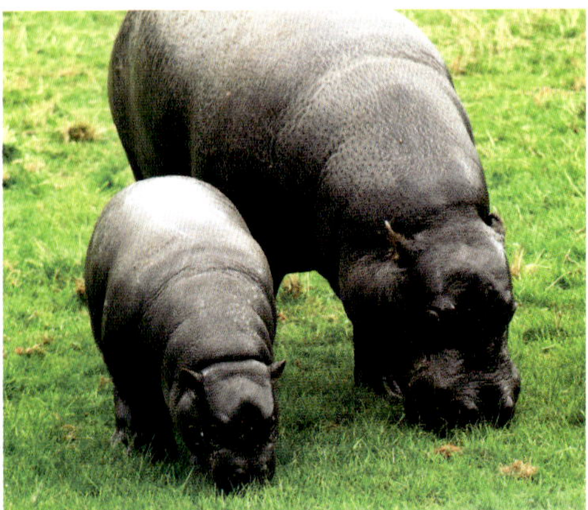

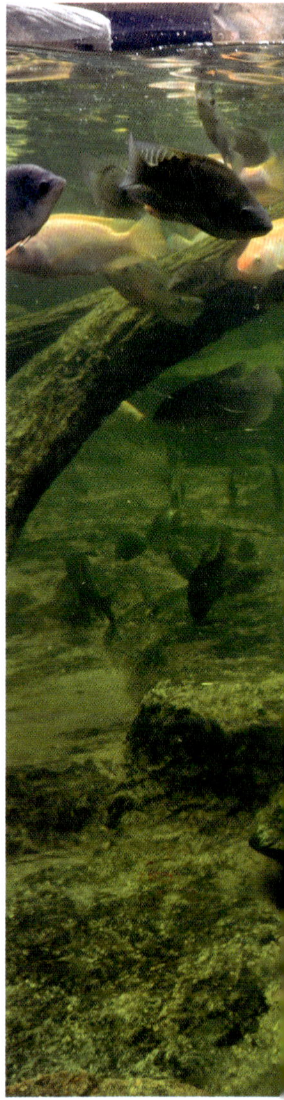

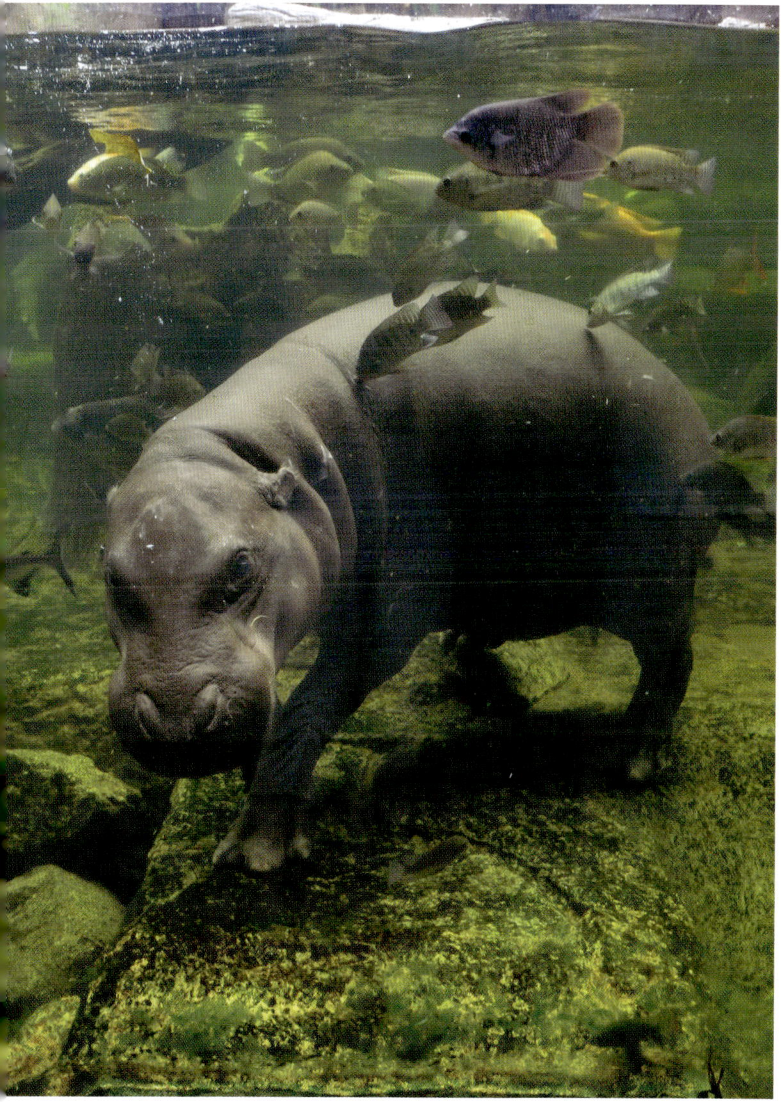

Giraffe

The tallest animal on Earth has a complex classification, with some authorities preferring to see the many separate populations being different species, some more endangered than others. Nevertheless, taken together, this iconic animal is still vulnerable to extinction due to habitat loss and poaching.

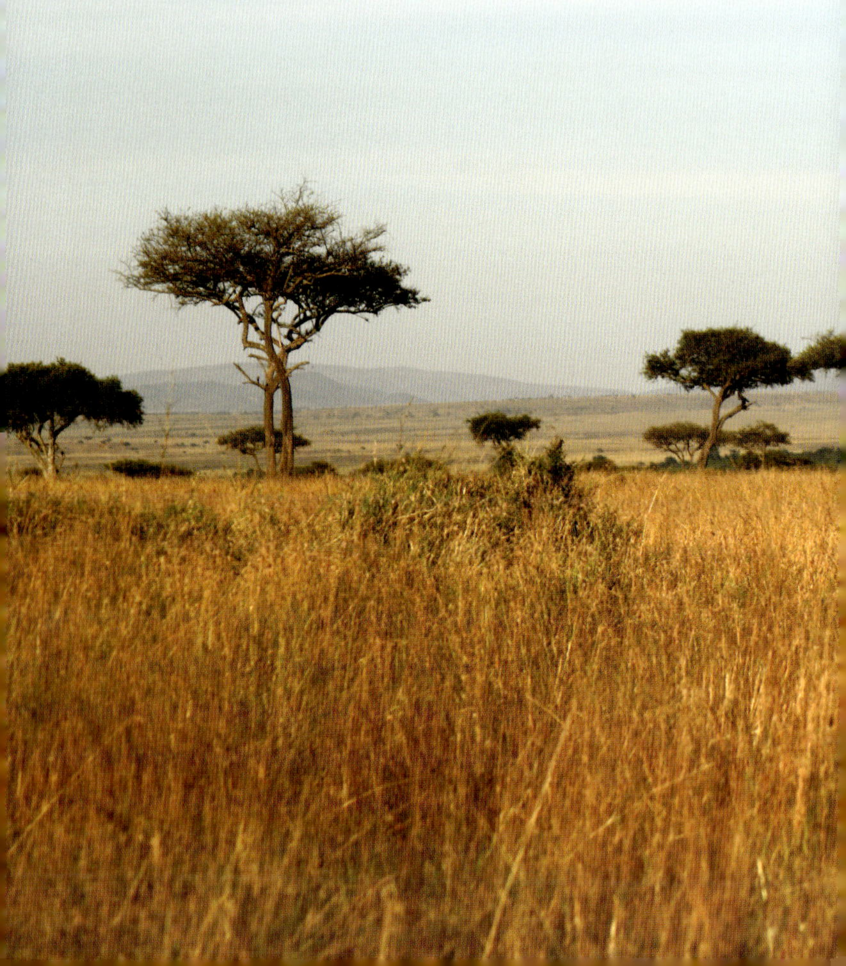

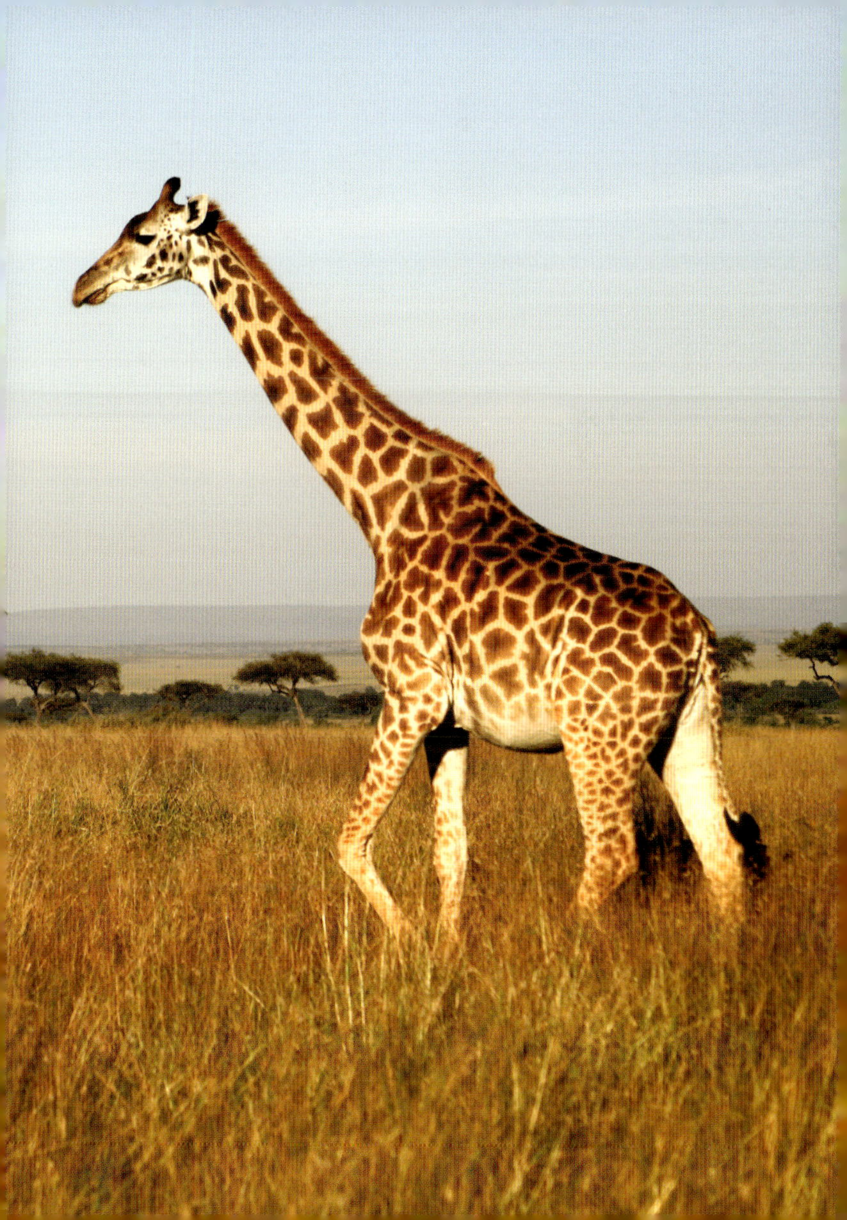

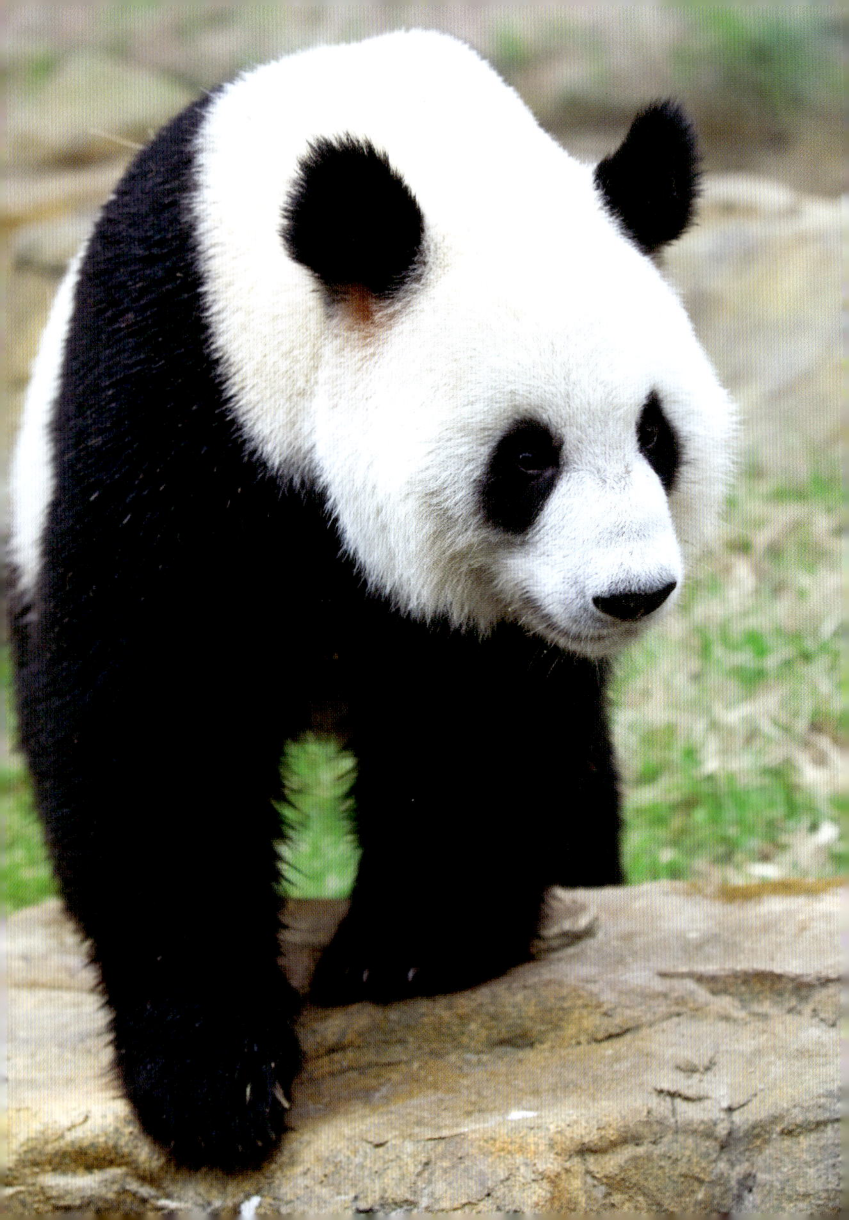

Asia and Australia

Asia is by far the largest continent, constituting a third of all the world's land. It spreads from the humid jungles around the equator, to the barren icy tundra of the High Arctic. From Japan in the east to the Mediterranean Sea in the west it reaches more than a quarter of the way around the planet. It is little surprise therefore that Asia plays host to a startling array of animals, great and small, and sadly many are struggling to survive against the onslaught of climate change, habitat destruction and hunting. The most famous Asian animals, the tiger, orang-utan and panda, are just three of many thousands listed as endangered.

In marked contrast with Asia, Australia is the smallest of the continents and has fewer than 40 million people living there compared the to 4.5 billion in Asia. Cut off from the rest of the world, Australia forms an arid paradise for a unique community of wildlife, most famous for its marsupial mammals, such as kangaroos and wombats. However, the arrival of colonisers and settlers from far-flung places in the last few centuries has devastated parts of the continent's natural heritage. As a result, 29 Australian mammal species have become extinct in the last 200 years, and dozens more of them are highly endangered today.

OPPOSITE:

Giant panda

For a long time, the panda's distinctive black-and-white coat was a global brand for endangered animals. After decades of effort, these incredible Chinese bears are now safe from extinction.

RIGHT TOP:

Giant panda

The giant panda is a singular off-shoot of the bear family that lives only in the steep mountain forests of southern China. By the end of the 20th century there were perhaps just 3,000 living wild.

RIGHT BOTTOM:

Panda cub

The future of pandas has been secured by breeding cubs in captivity – often using in vitro fertilization techniques – and then reintroducing them to the wild.

OPPOSITE:

Bamboo diet

Pandas eat only the leaves and stalks of bamboo that grow in dense glades in their forest habitat. Pandas are unable to survive on any other foods.

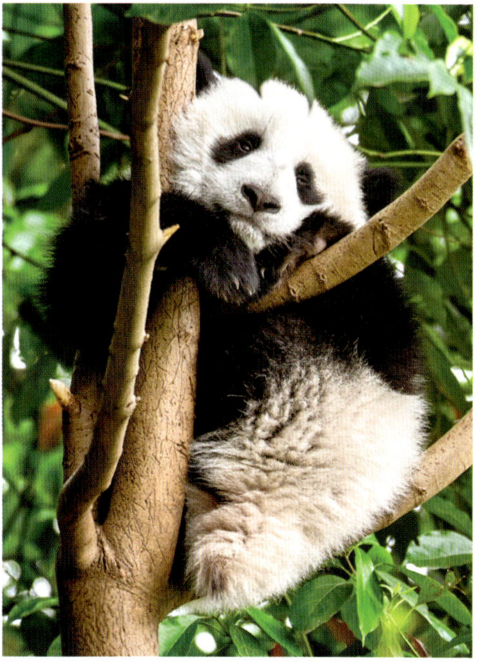

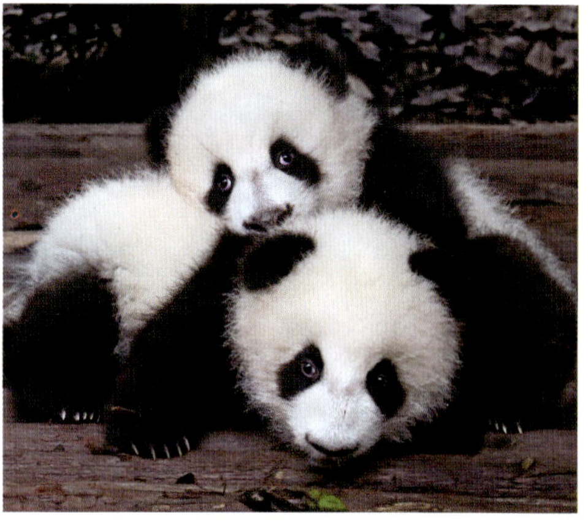

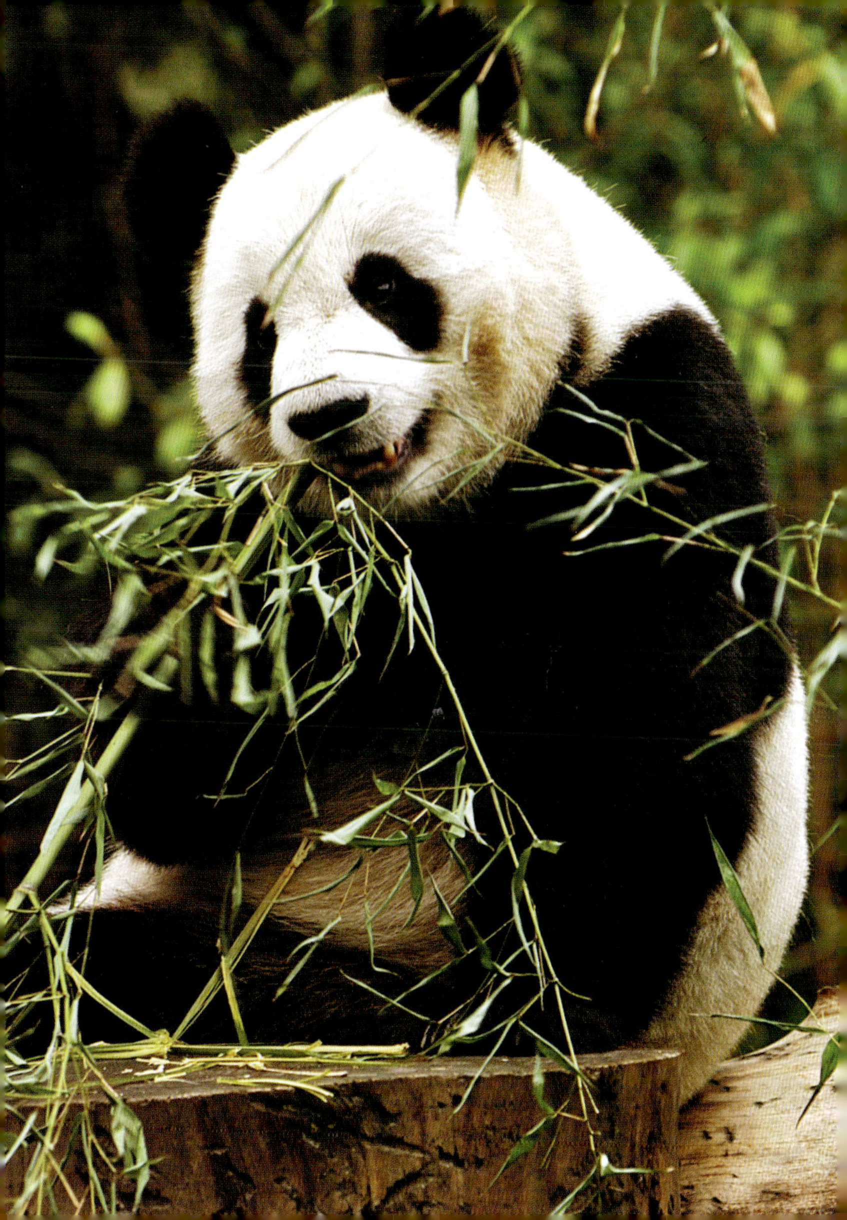

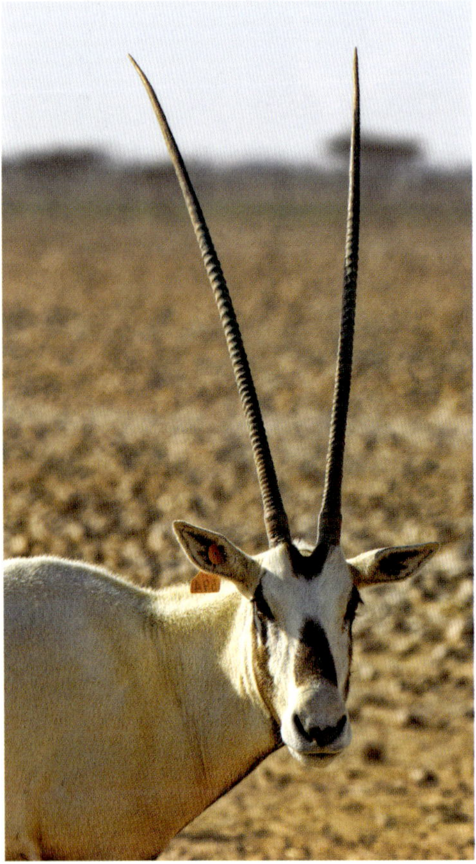

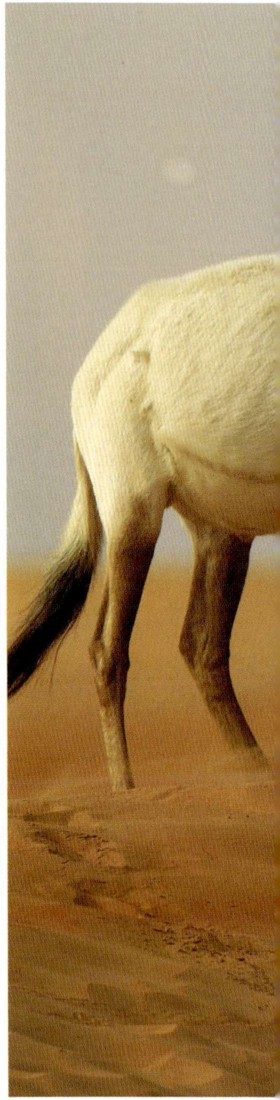

ALL PHOTOGRAPHS:

Arabian oryx

Thanks to sports hunting using off-road vehicles and high-tech rifles these desert antelopes became extinct in the wild in the 1970s. However, an intensive programme of captive breeding, dubbed Operation Oryx, means there are now 2,200 wild oryx today, and many more in zoos and nature parks around the world.

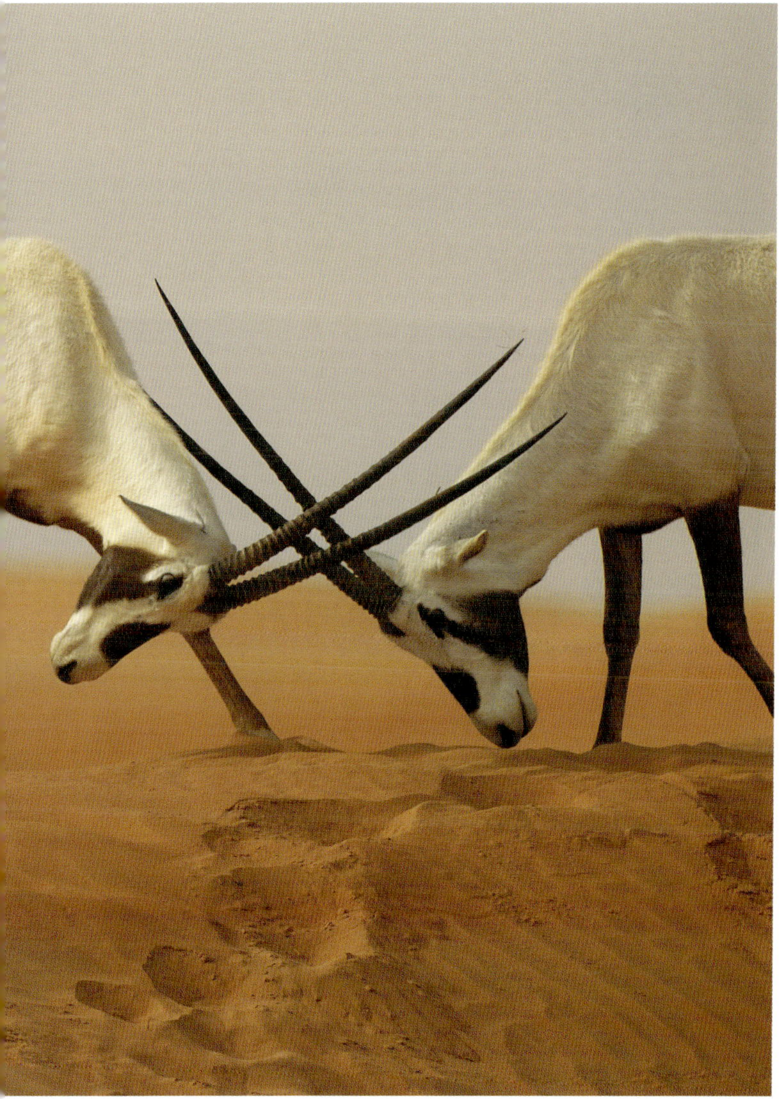

Asian elephant
Unlike African elephants, their smaller cousins over in south and east Asia have been widely domesticated as a beast of burden. However, the number of truly wild Asian elephants is around 50,000 and falling.

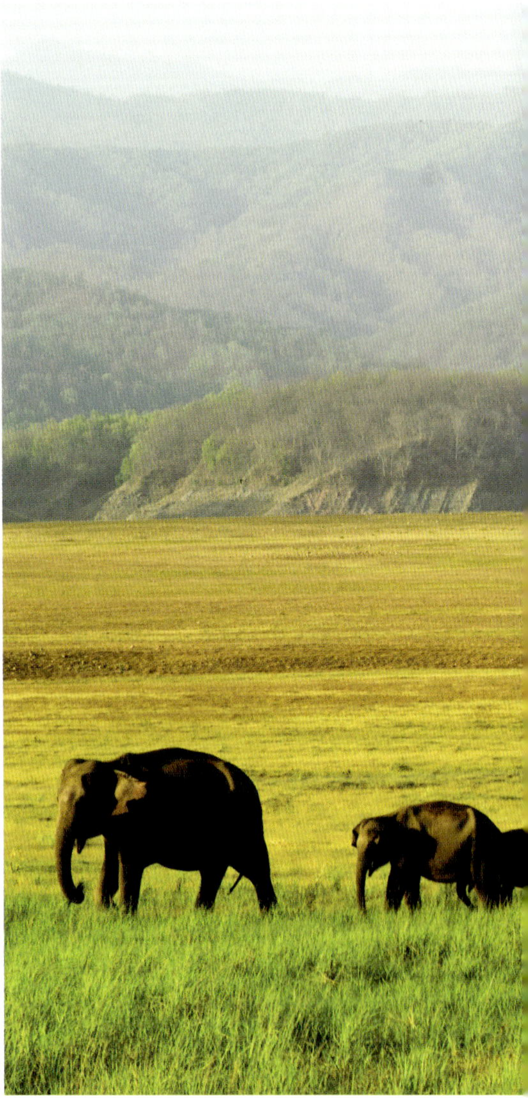

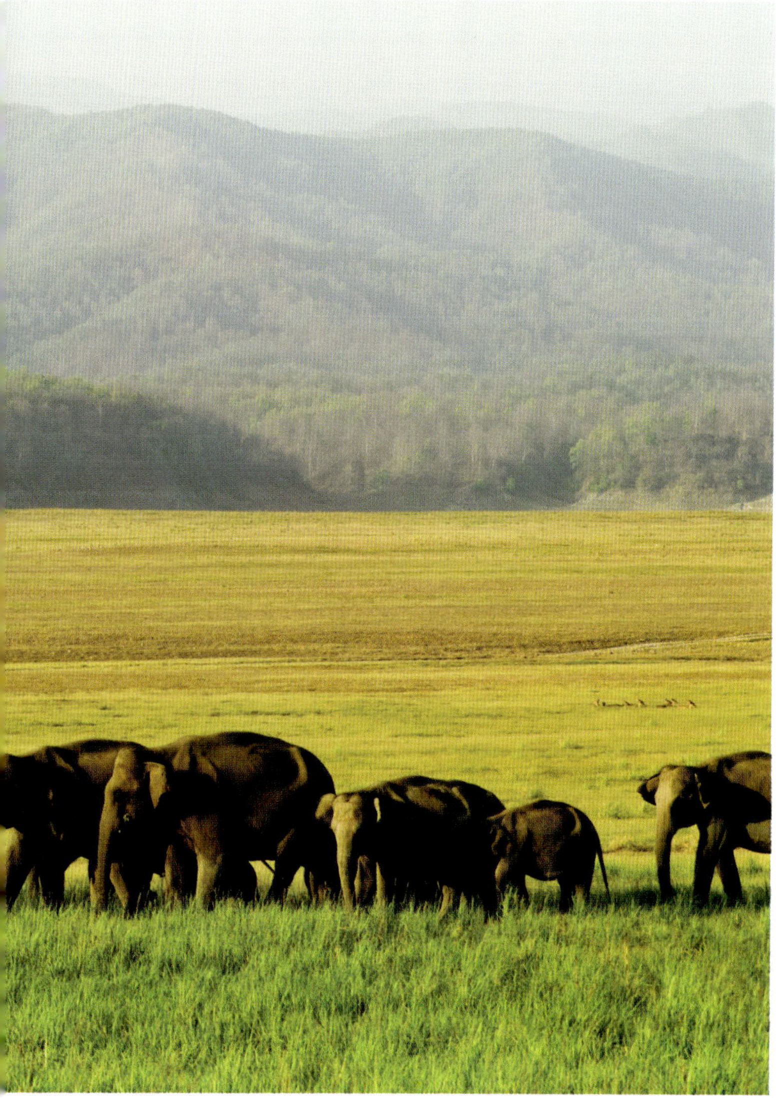

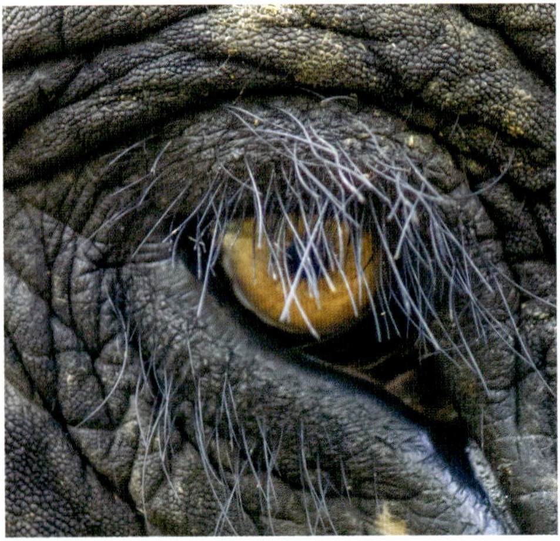

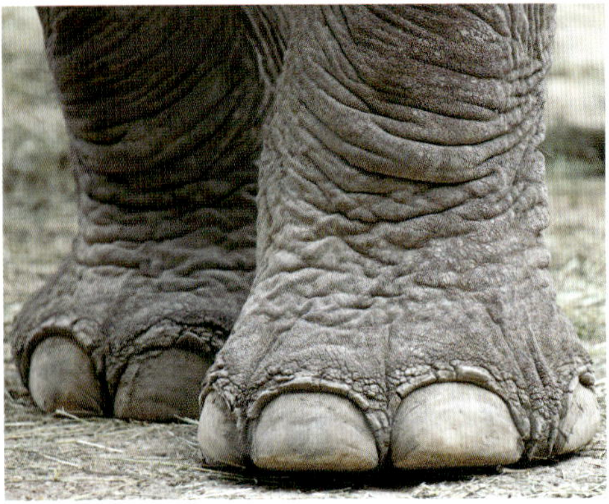

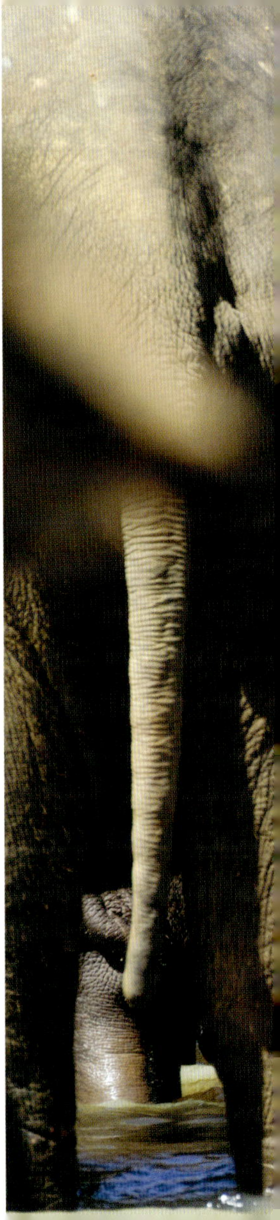

ALL PHOTOGRAPHS:

Asian elephant

The Asian elephant's historical range spreads from the eastern Mediterranean to eastern China, but today it is mostly confined to fragments of jungles in India, Southeast Asia and Indonesia. As their habitat shrinks the giant beasts are frequently in conflict with humans and always come out worse in these encounters.

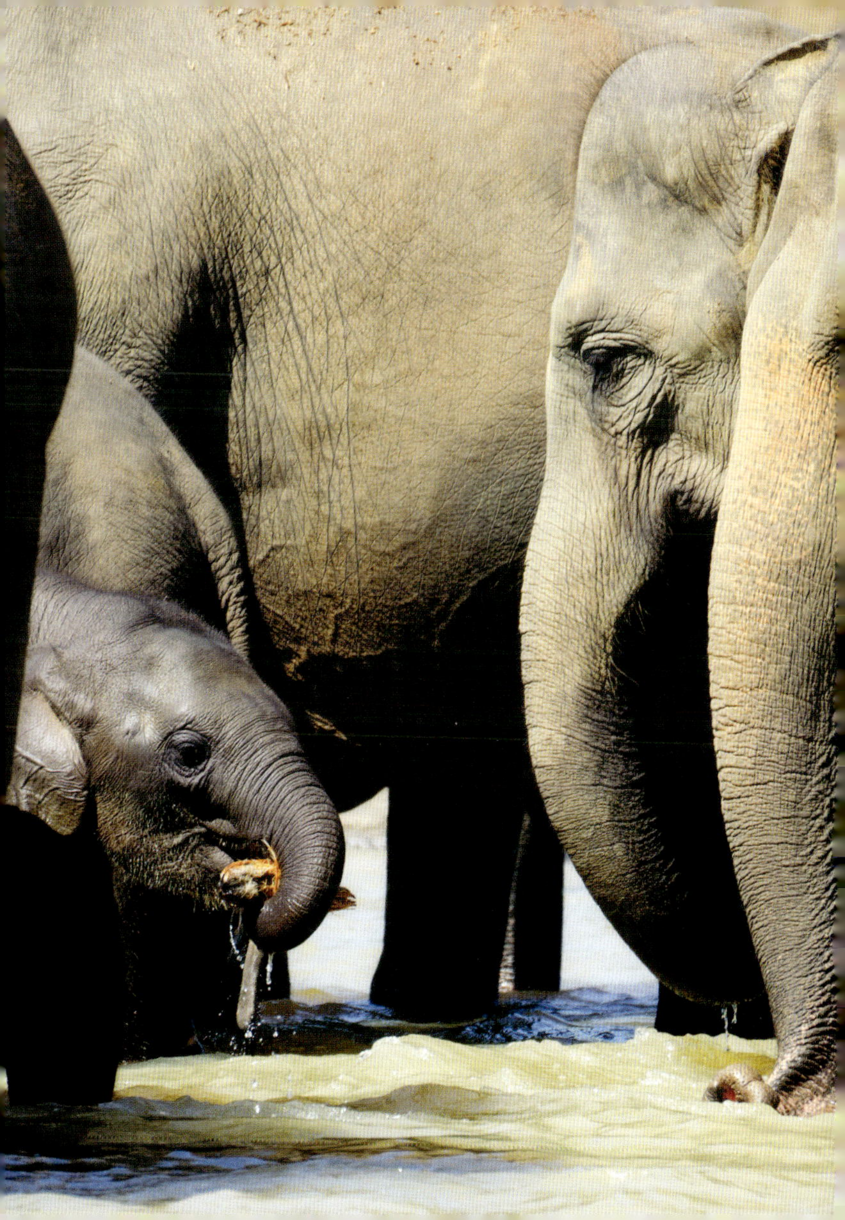

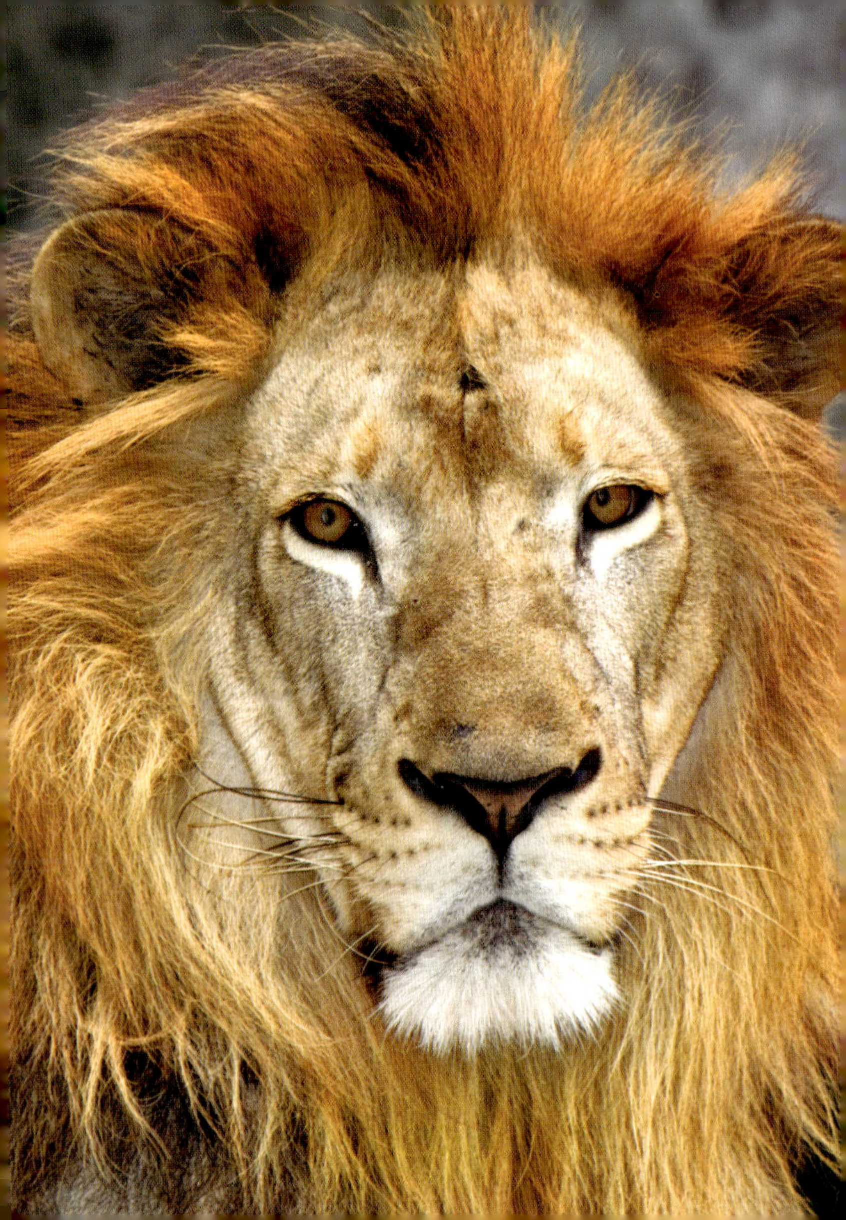

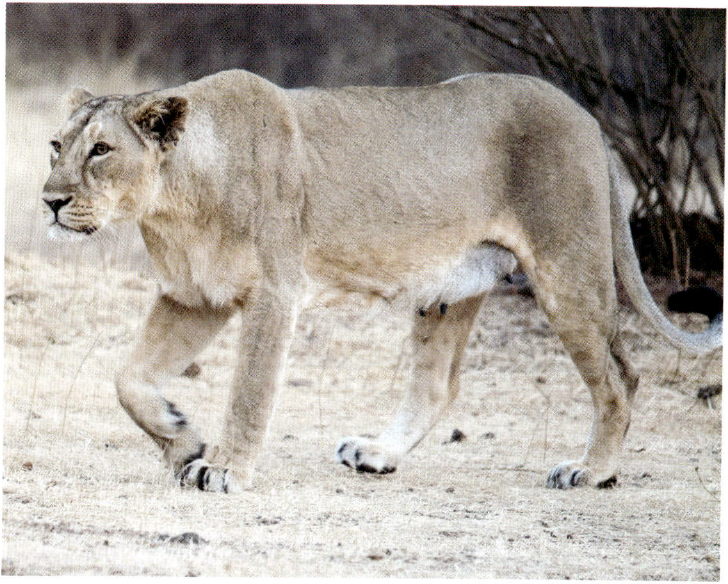

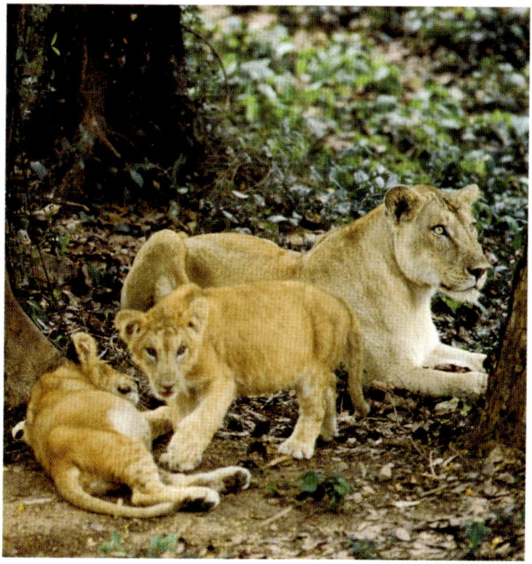

ALL PHOTOGRAPHS:

Asiatic lion

These lions are confined to the Gir Forest in western India. They are almost identical in size and shape – and lifestyle – to the African species. As of 2010, there were about 500 wild lions, a marked increase from near extinction just a few years before.

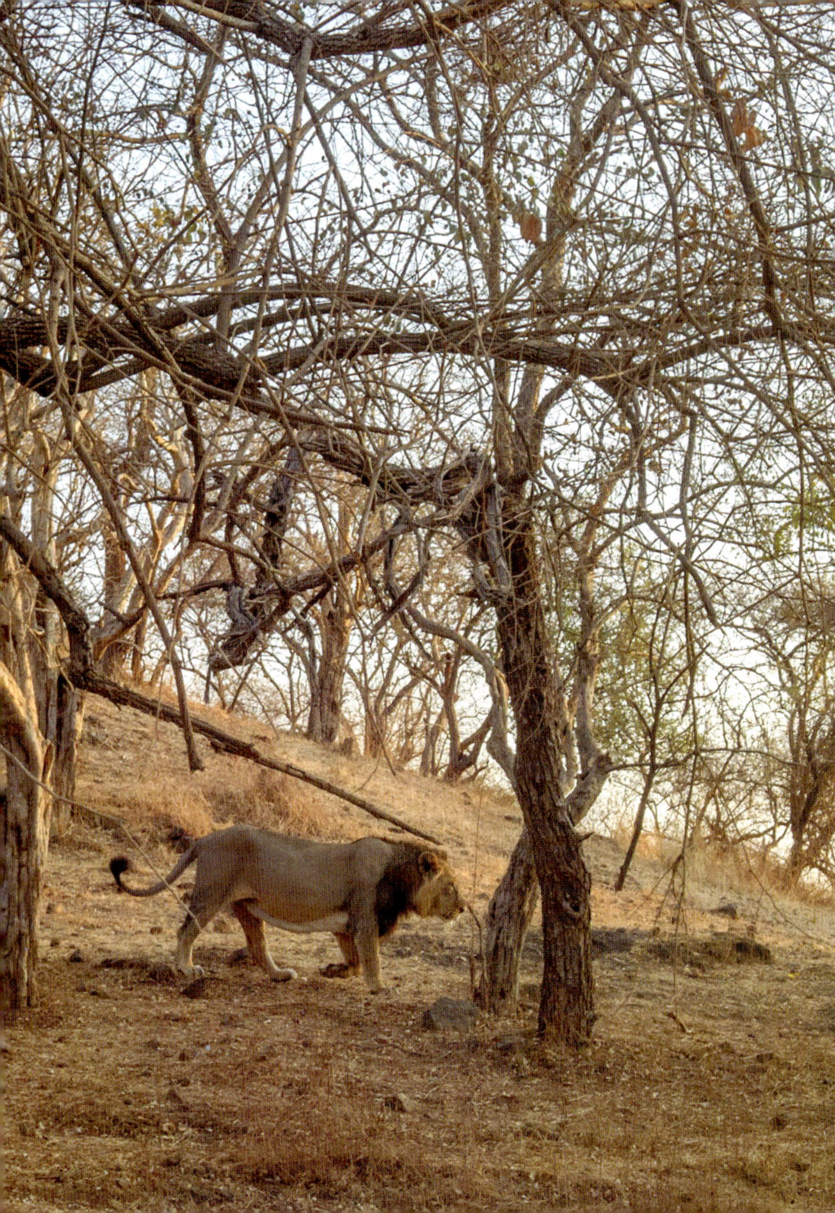

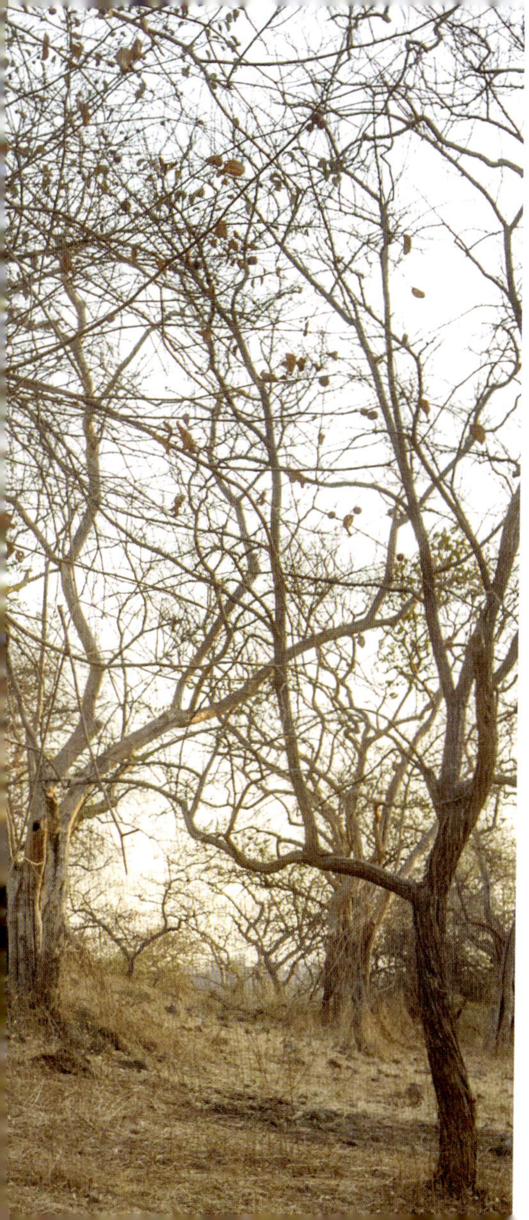

Asiatic lion
One defining feature of Asian lions is that the manes of the males rarely extend far on to the neck and shoulders, unlike their African counterparts.

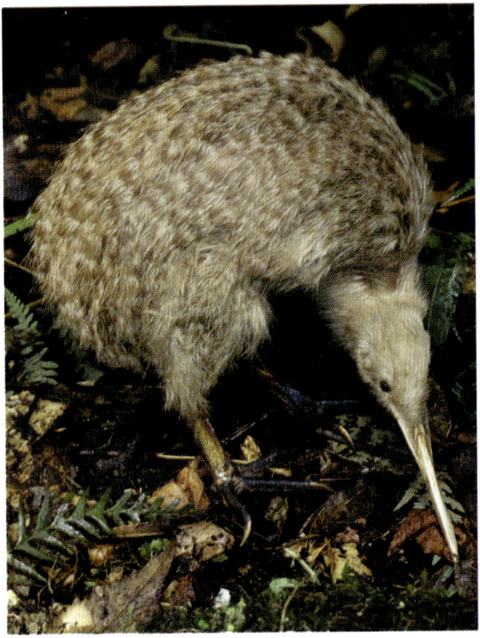

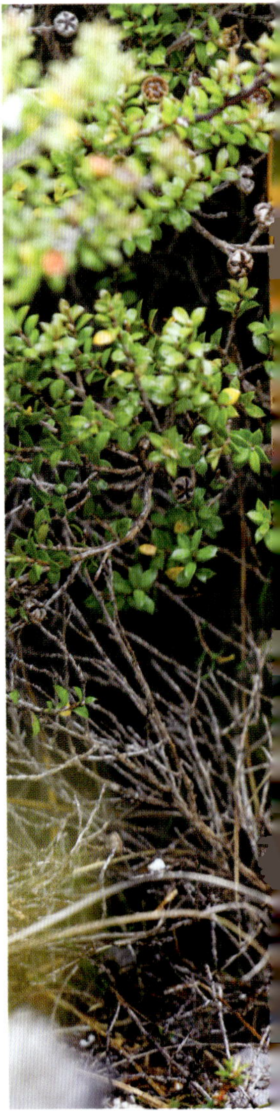

ABOVE:
Kiwi
Found only in New Zealand, the flightless kiwi plunges its long beak into soft soil and sand to grab worms and insects. Four out of five of the kiwi species are endangered, due in part to habitat loss and being over-hunted for their distinctive feathers. Additionally, mammals introduced to New Zealand prey on the birds and their nests.

RIGHT:
Kākāpō
This very unusual parrot from New Zealand is one of rarest birds on Earth, numbering around 200. The bird is flightless and nocturnal (and produces the loudest call of any bird) but is highly vulnerable to attack from mammal predators.

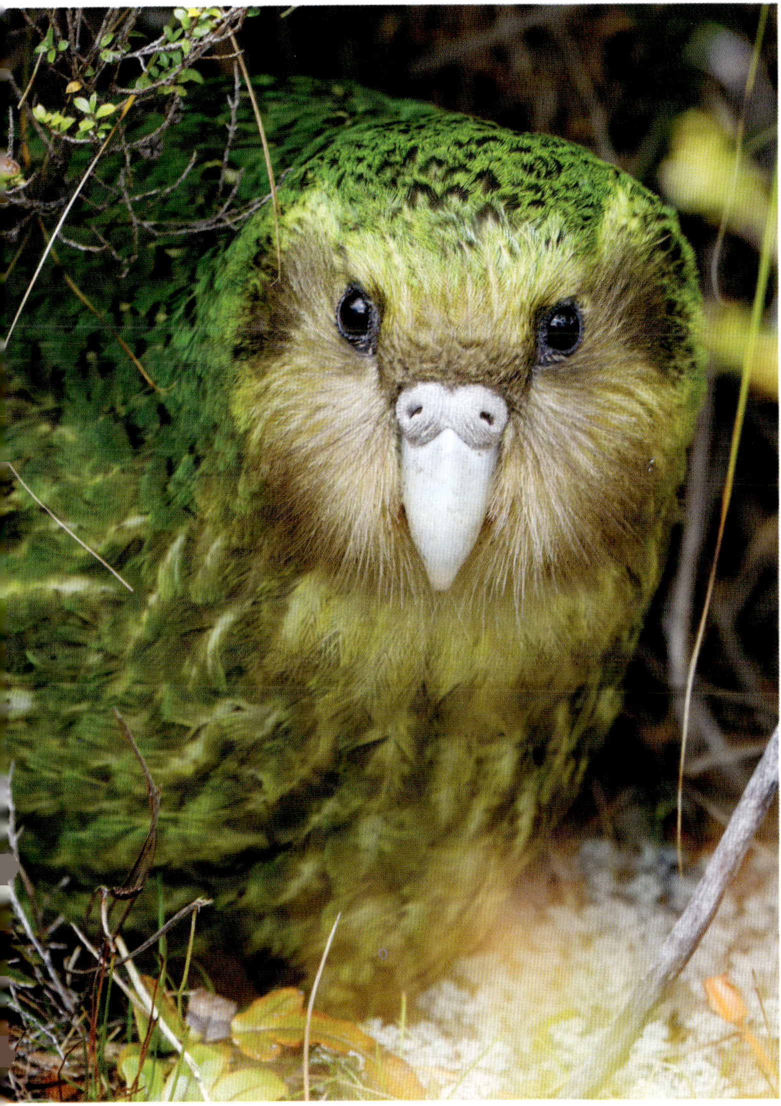

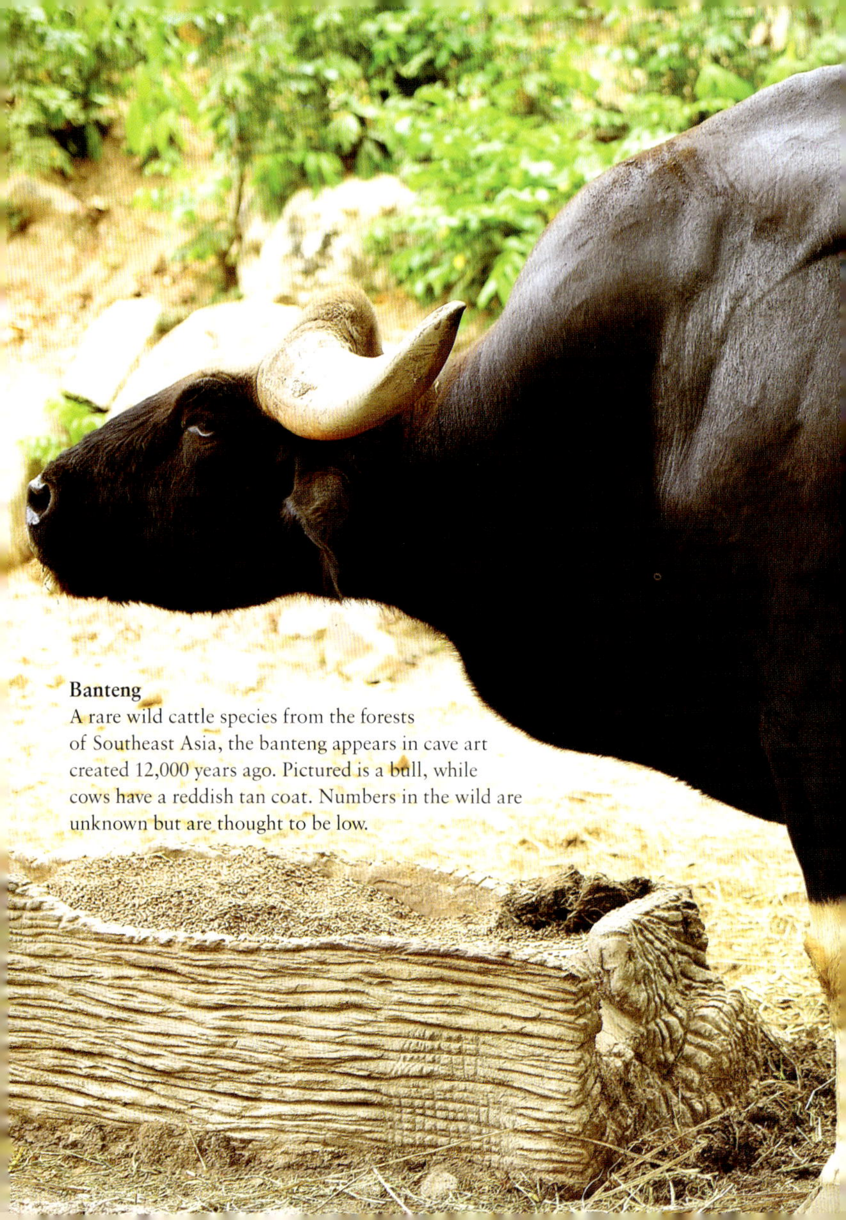

Banteng
A rare wild cattle species from the forests of Southeast Asia, the banteng appears in cave art created 12,000 years ago. Pictured is a bull, while cows have a reddish tan coat. Numbers in the wild are unknown but are thought to be low.

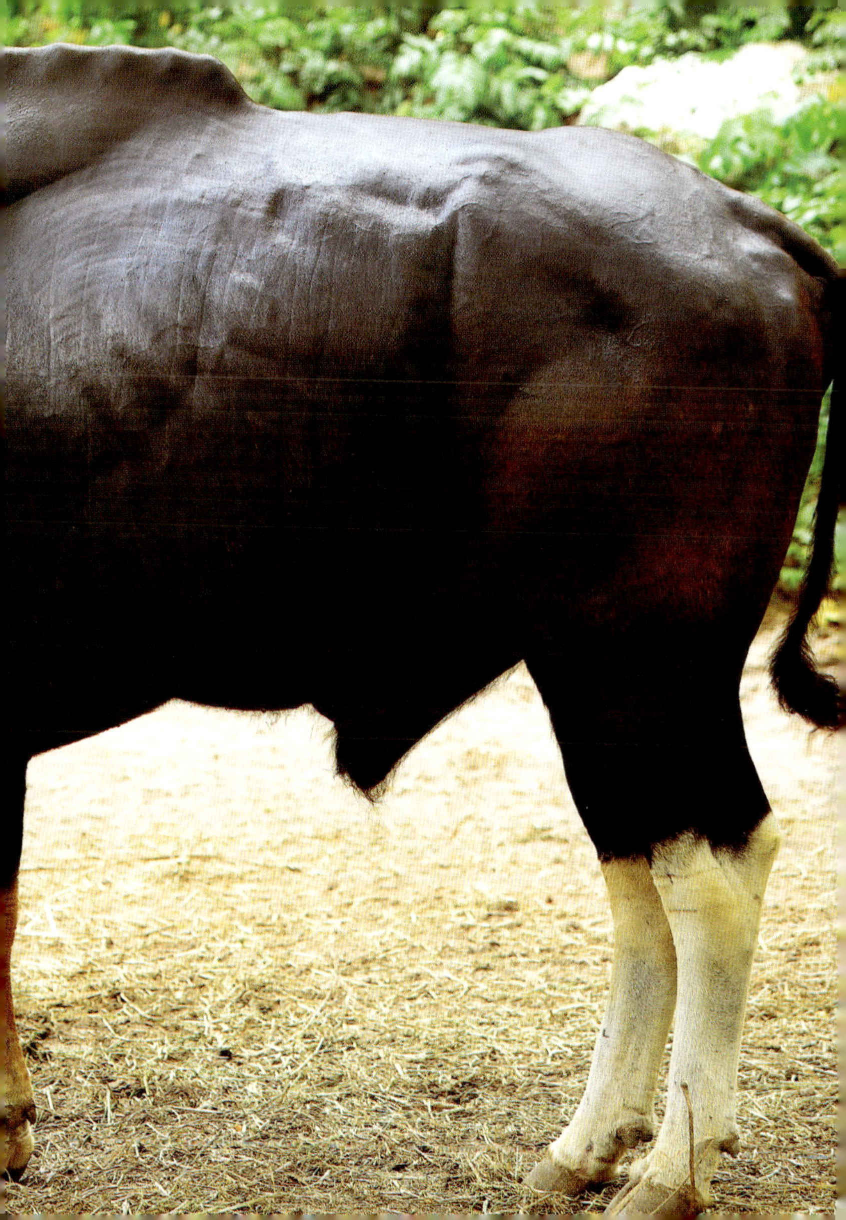

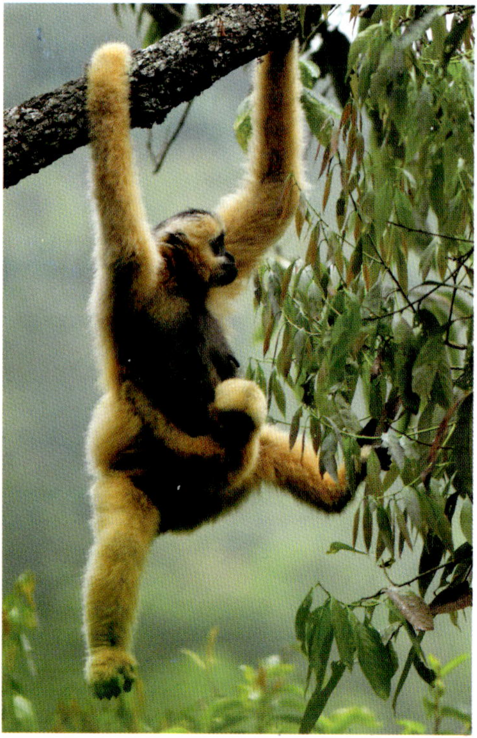

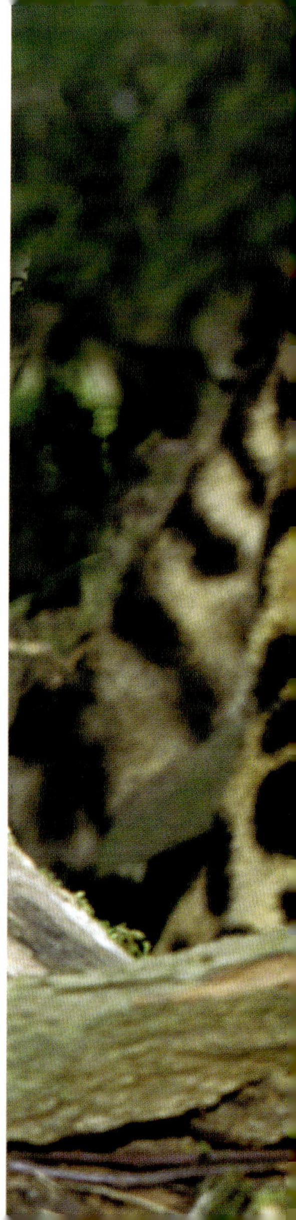

ABOVE:

Black-crested gibbon

The gibbons (or lesser apes) all live in and around the forests of Southeast Asia. There are 18 species, including this one from the north of Indochina, and every one is under threat from extinction, often critically so. The gibbons, which swing high in the branches, are suffering due to deforestation.

RIGHT:

Clouded leopard

This is the smallest of the big cats, which roar instead of purr. It lives in the forests of East Asia and is hunted for its mottled coat.

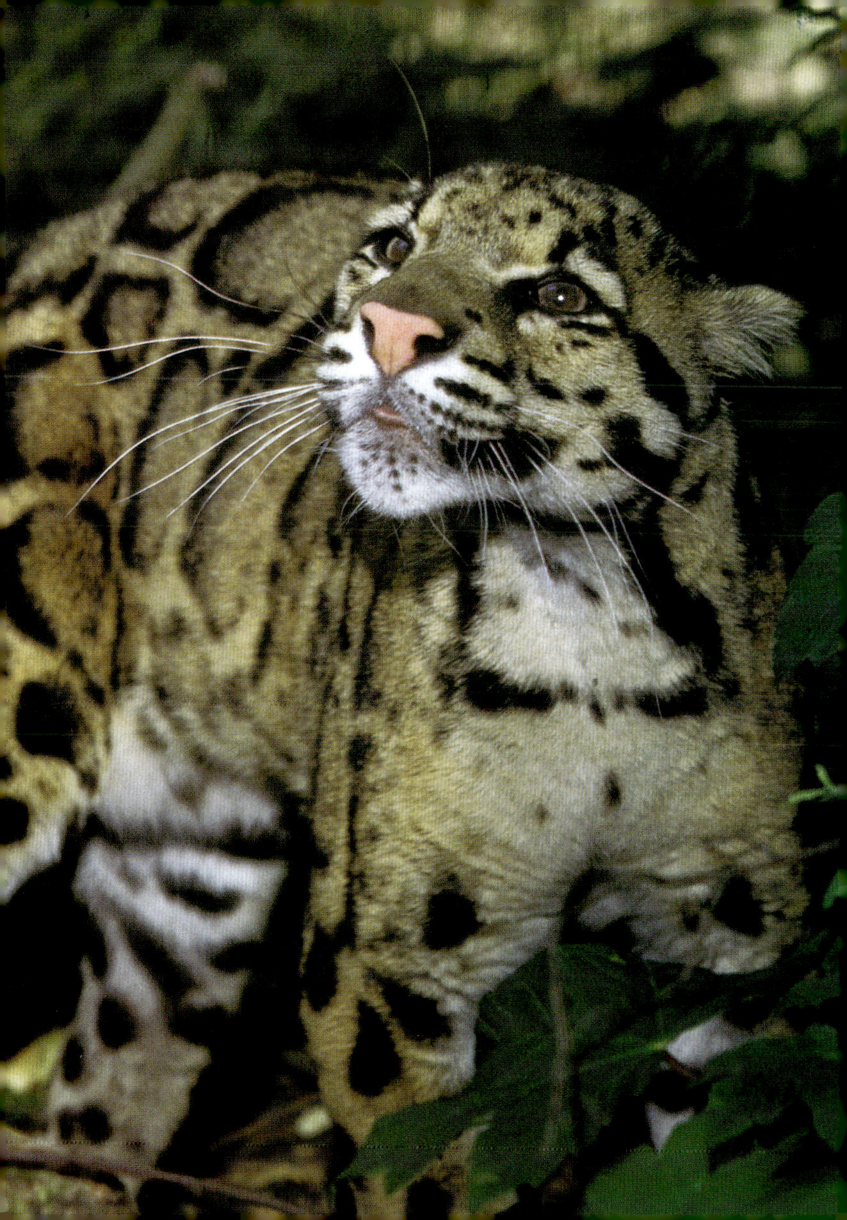

RIGHT:
Chinese crocodile lizard
Named after the way its rough skin gives it a passing resemblance to a crocodile, this semi-aquatic lizard is classed as vulnerable to extinction. The main threats it faces include habitat loss and being killed by human river fishers.

OVERLEAF LEFT:
Dhole
Also known as the Asia red dog, the dhole is a pack hunter from the forests of Central and South Asia. Although it seldom attacks livestock, the dogs are traditionally persecuted across their range.

OVERLEAF RIGHT:
Double-wattled cassowary
This is one of the three flightless cassowary species in the forest of New Guinea that requires protection from deforestation and hunting.

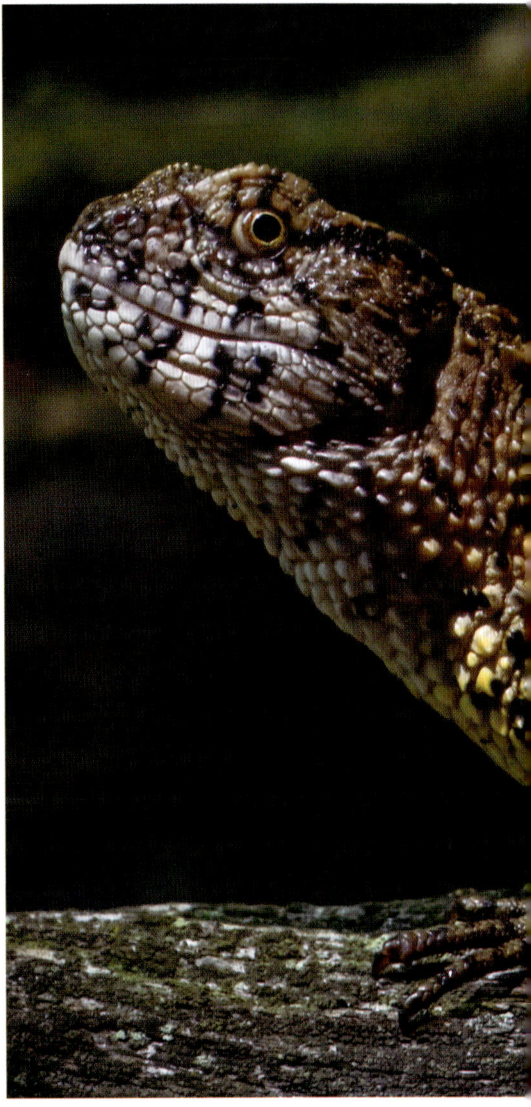

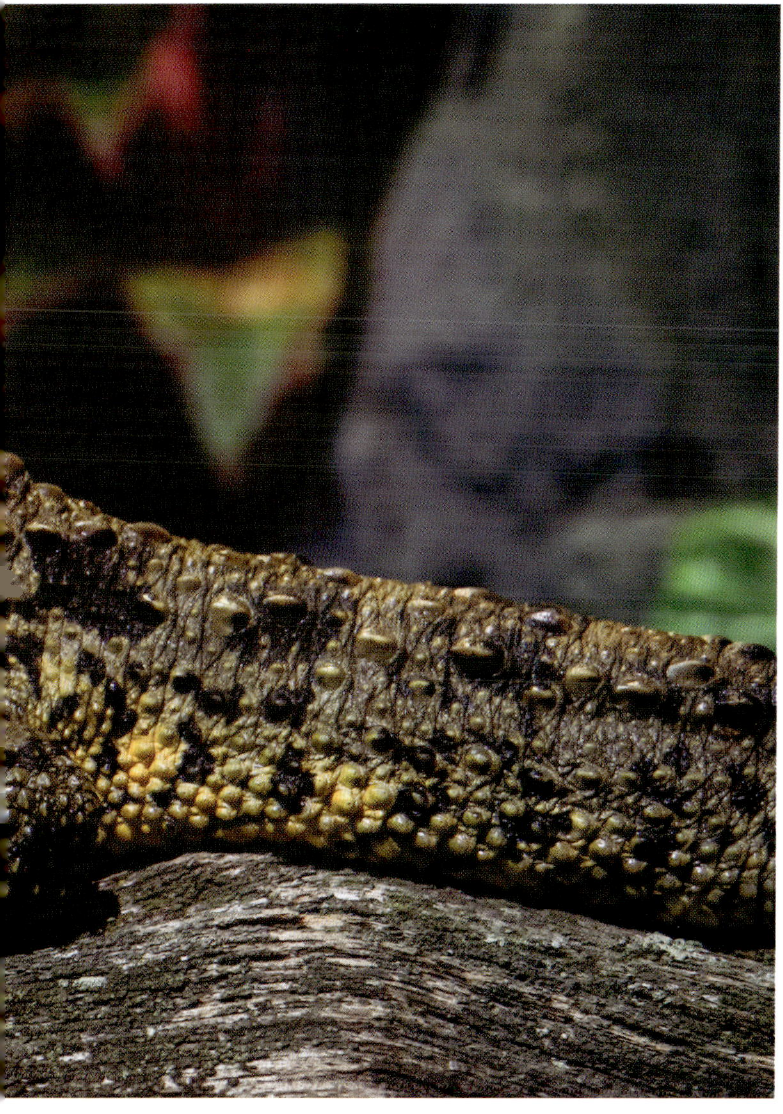

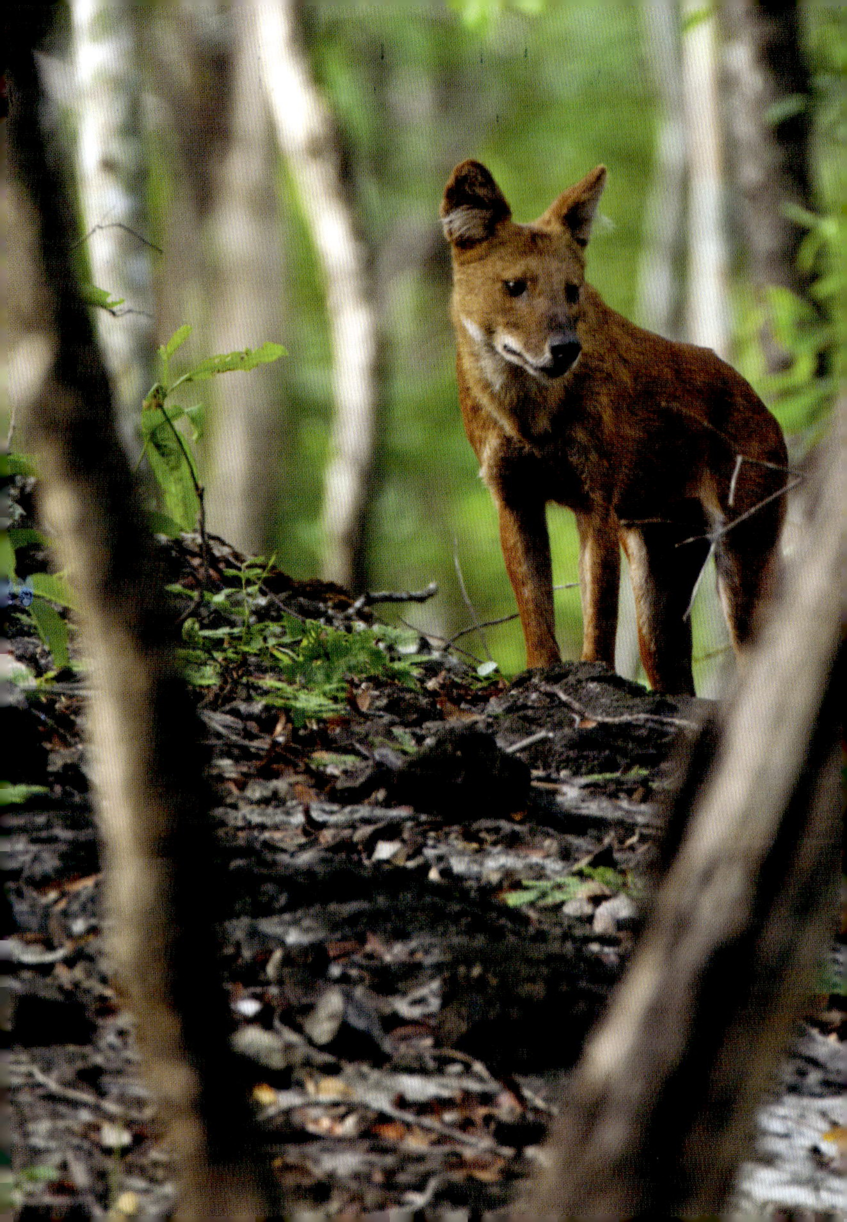

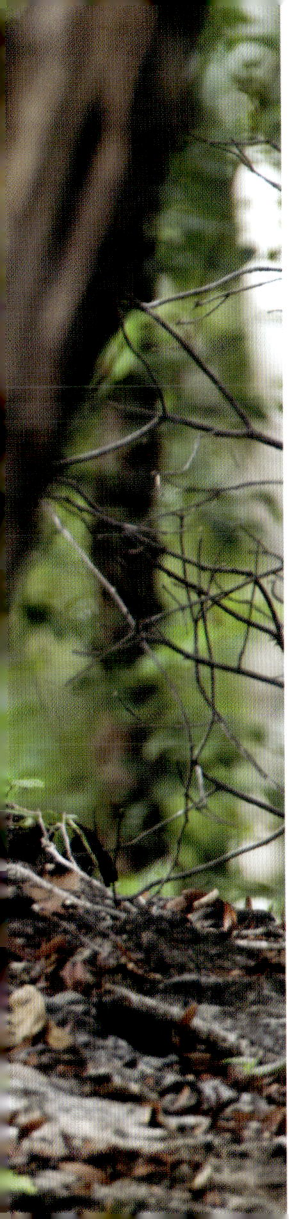

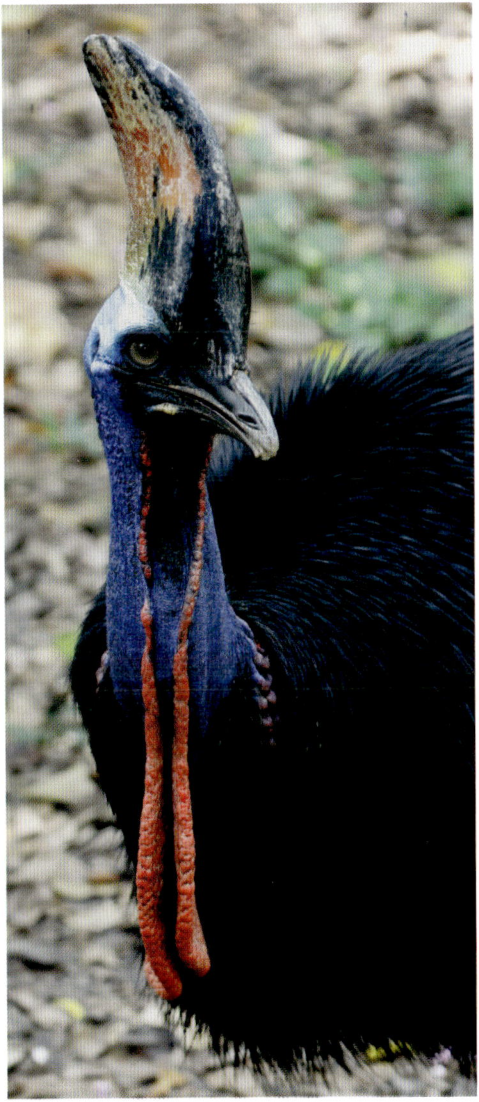

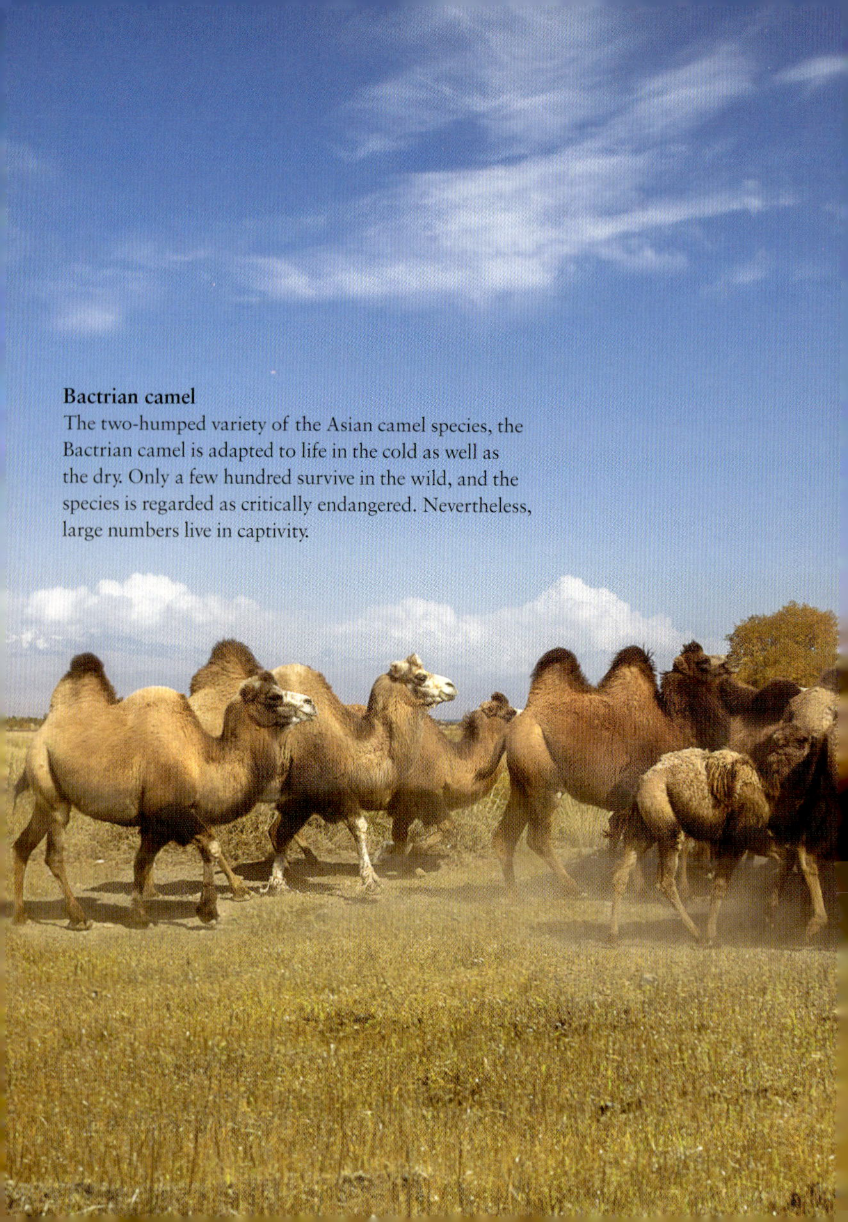

Bactrian camel

The two-humped variety of the Asian camel species, the Bactrian camel is adapted to life in the cold as well as the dry. Only a few hundred survive in the wild, and the species is regarded as critically endangered. Nevertheless, large numbers live in captivity.

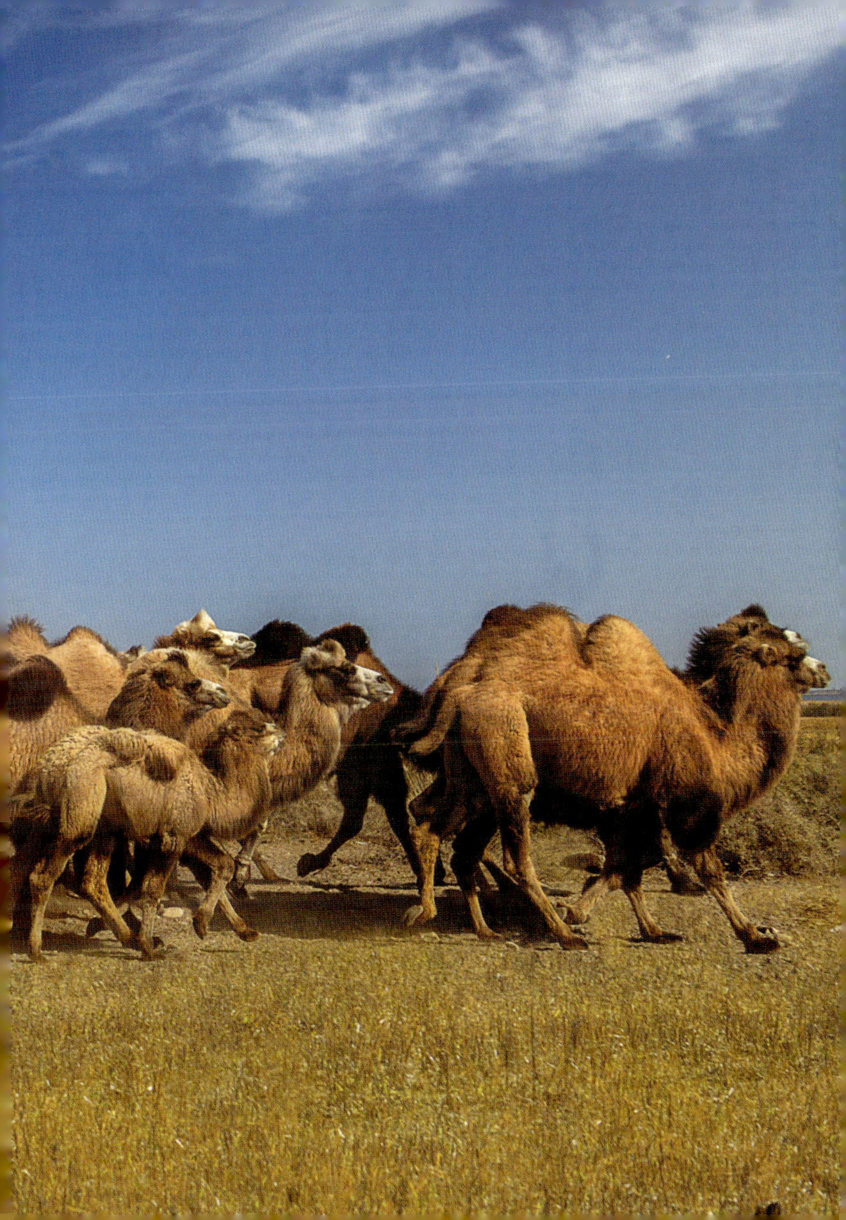

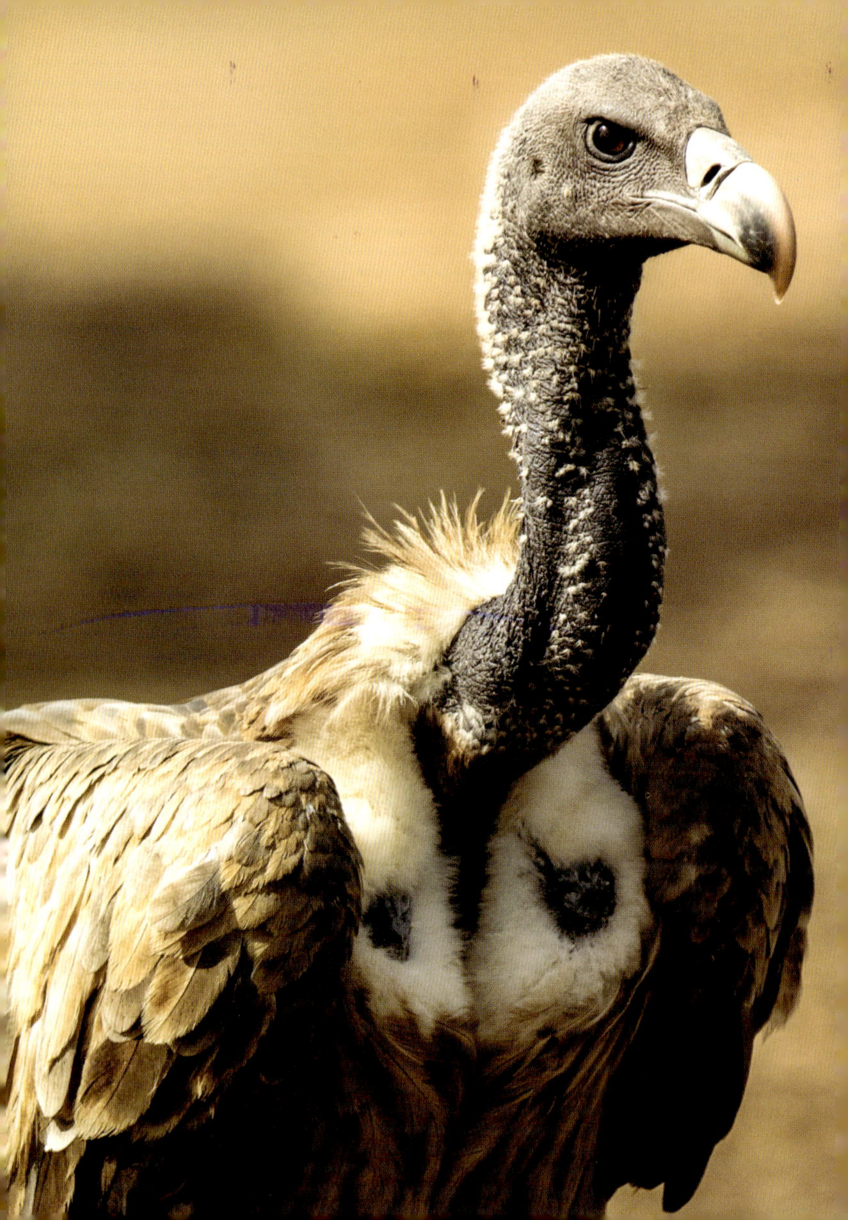

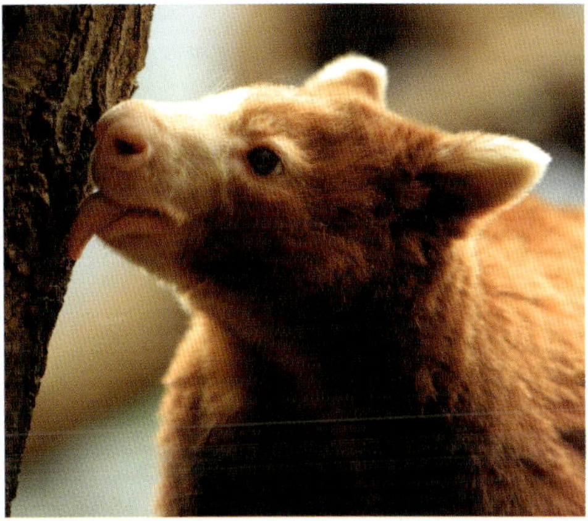

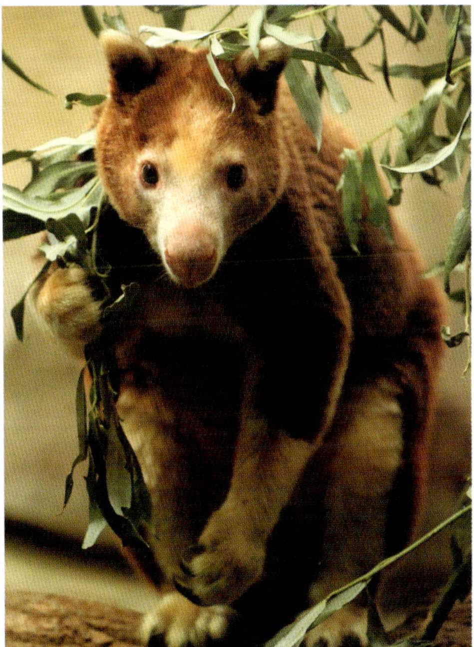

OPPOSITE:

Indian vulture

Once a significant scavenger in South Asia, the Indian vulture is now critically endangered due to a calamitous misuse of drugs given to oxen and other beasts of burden that allowed them to work harder. Once these animals die, the drug pass to the vultures that scavenge on their remains, leading to kidney failure and death. Today the species is being supported by captive breeding until this problem is resolved.

LEFT TOP AND BOTTOM:

Huon tree kangaroo

The tree kangaroos live in the forests of New Guinea (as does this species) and Queensland. The little marsupials are all at risk from deforestation and hunting.

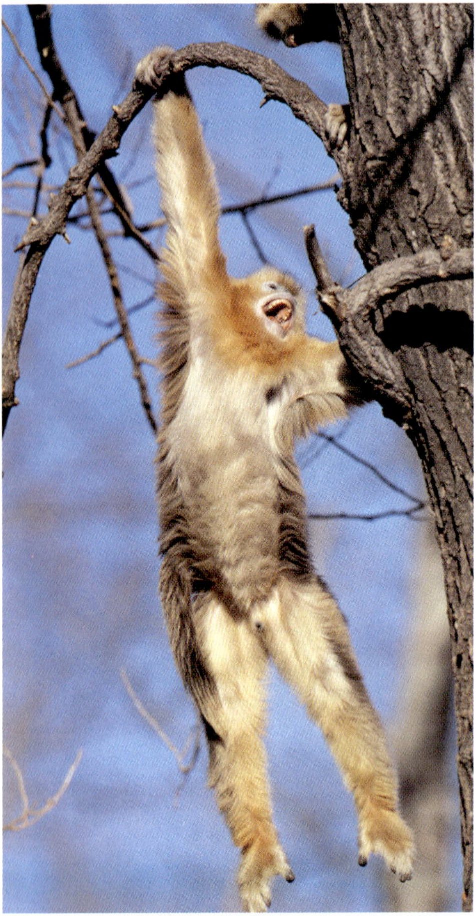

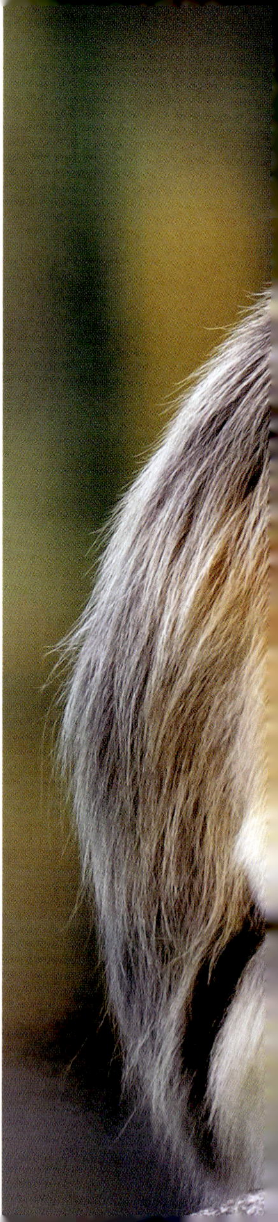

ALL PHOTOGRAPHS:

Golden snub-nosed monkey

This distinctive Chinese monkey is falling in number due to the degradation of its habitat. Its staple food is lichen that grows on dead trees in the forest, but these logs are often removed by people for firewood.

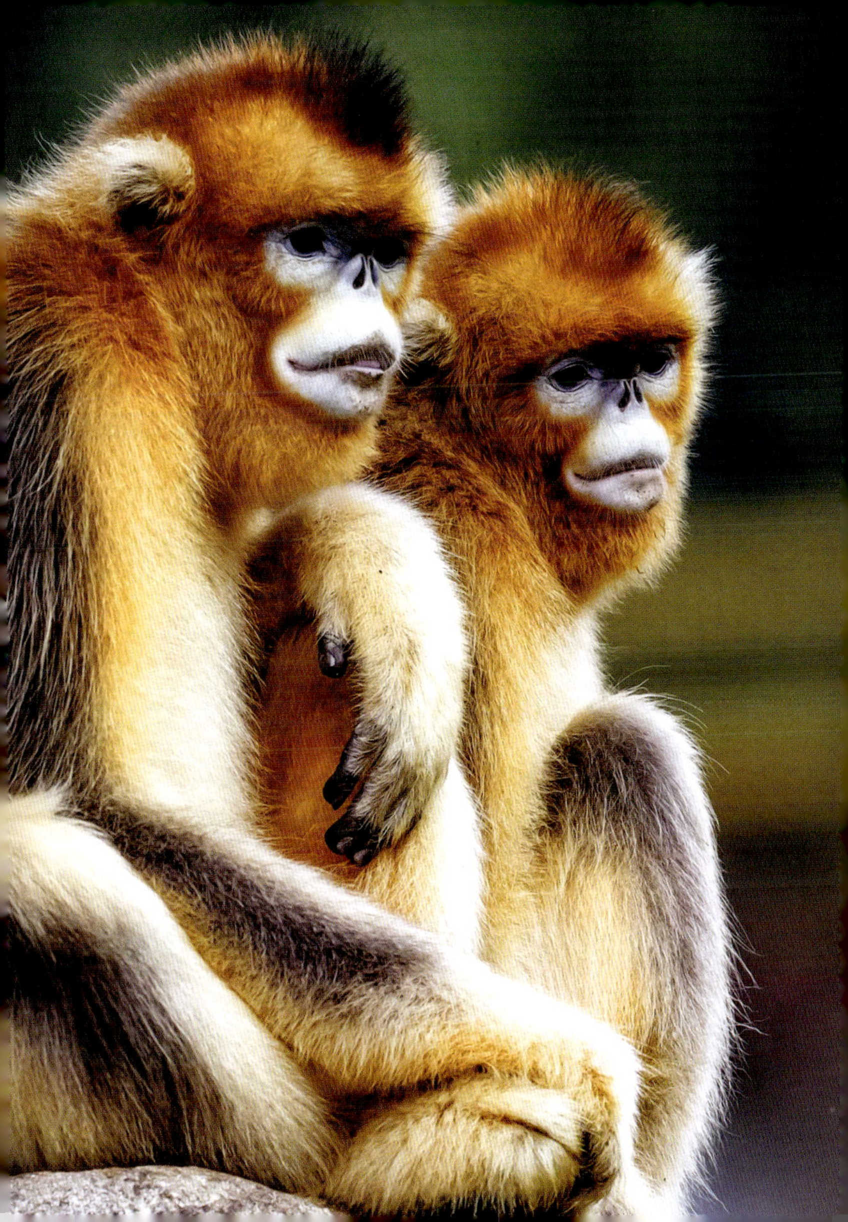

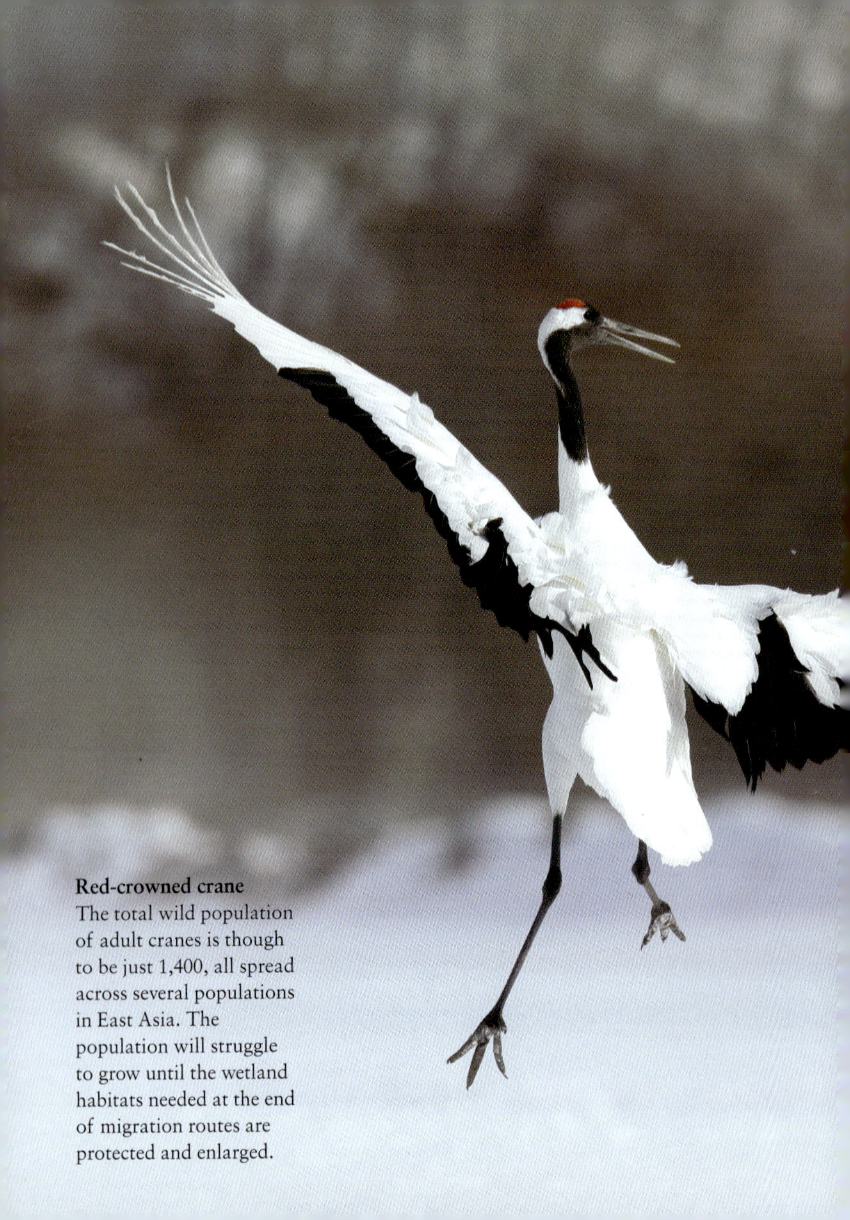

Red-crowned crane
The total wild population of adult cranes is though to be just 1,400, all spread across several populations in East Asia. The population will struggle to grow until the wetland habitats needed at the end of migration routes are protected and enlarged.

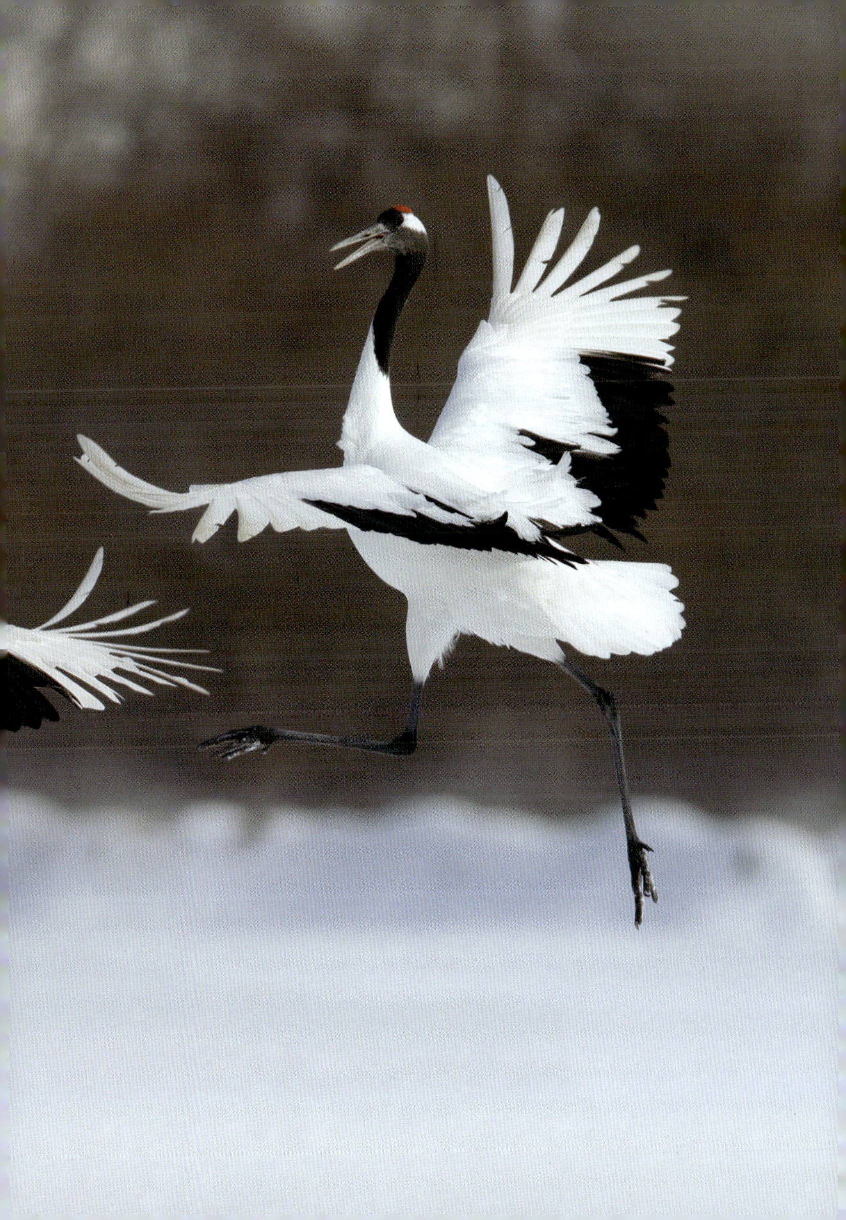

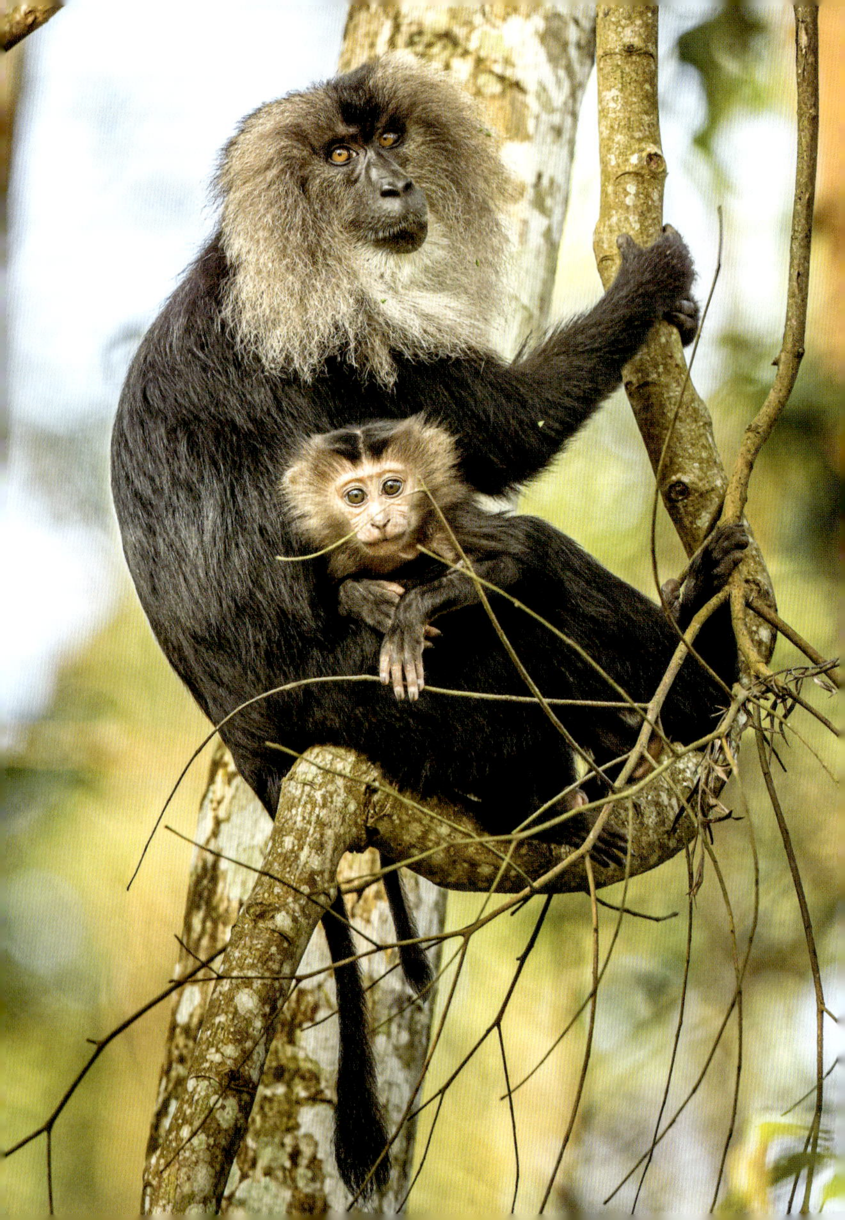

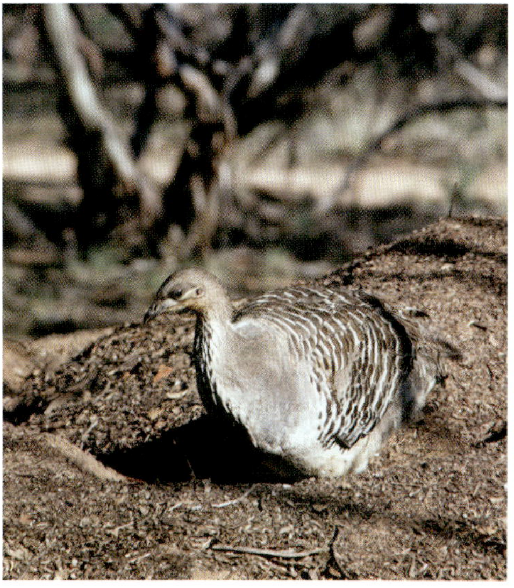

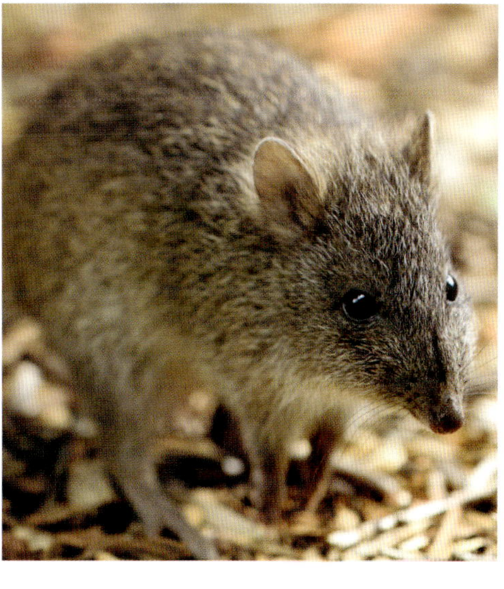

OPPOSITE:

Lion-tailed macaque

Restricted to the Indian coast of the Arabian Sea, this is one of the rarest monkey species, numbering 3,000 or so. They are threatened by agriculture destroying wild habitats.

LEFT TOP:

Malleefowl

A distant relative of the domestic chicken, this Australian bird buries its eggs in an earth mound and then leaves the chicks to their fate. Foxes introduced to Australia have decimated the malleefowl population and it is now classified as vulnerable.

LEFT BOTTOM:

Long-footed potoroo

In common with many marsupial species, the long-footed potoroo is threatened because it cannot withstand the duel onslaught of habitat loss and predation by mammals, such as foxes and cats, introduced from other parts of the world over the last few centuries.

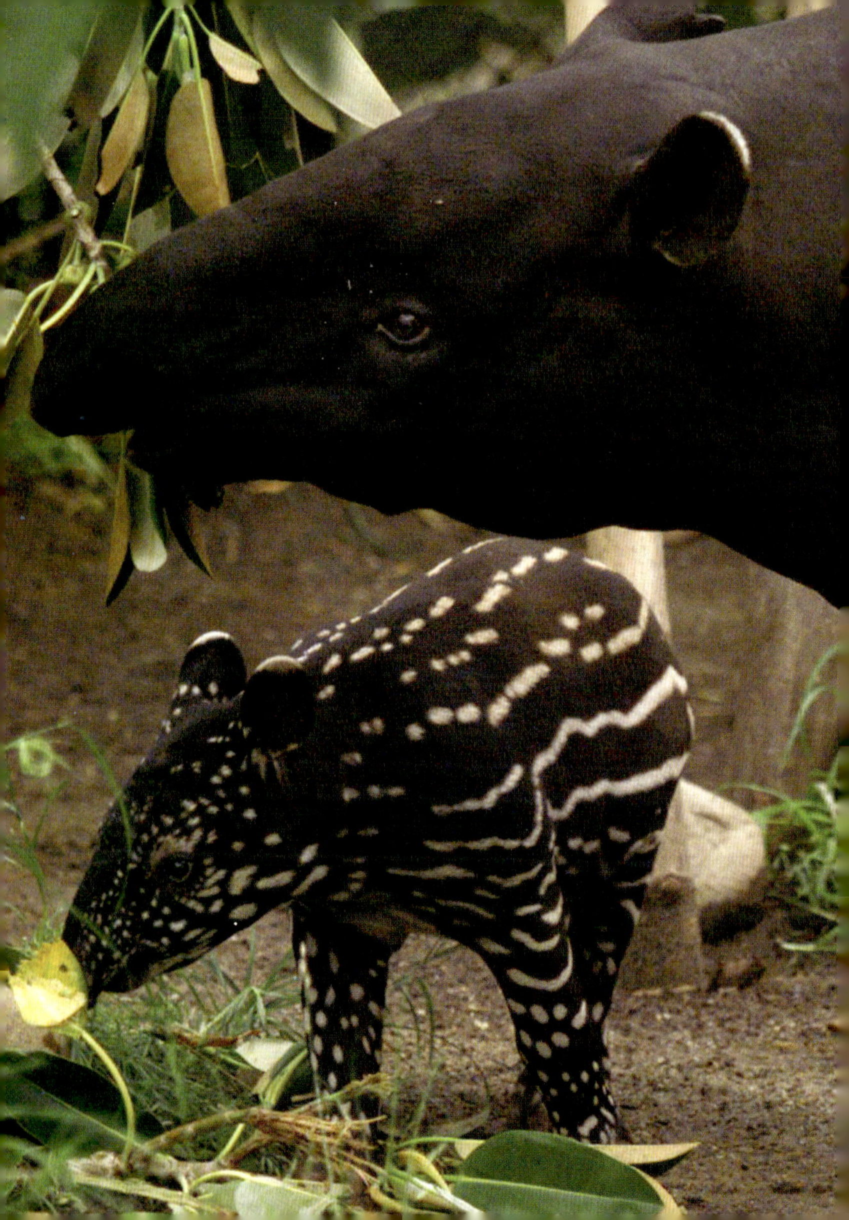

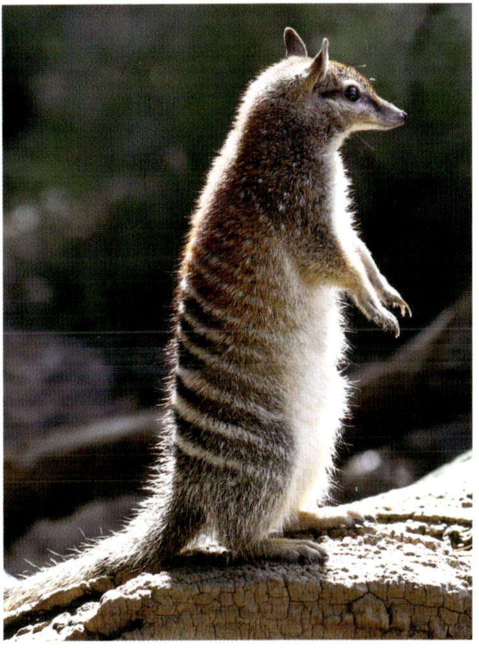

LEFT AND BELOW:

Numbat

This endangered marsupial uses its long tongue to slurp up termites. The animal has been wiped out in much of its range by the introduction of the red fox. Today they survive in areas where they can seek refuge from predators inside hollow logs.

OPPOSITE:

Malay tapir

A forest browser that uses its flexible lips to strip leaves from bushes is endangered across its Southeast Asian home as its habitat is steadily destroyed. They are also hunted as food.

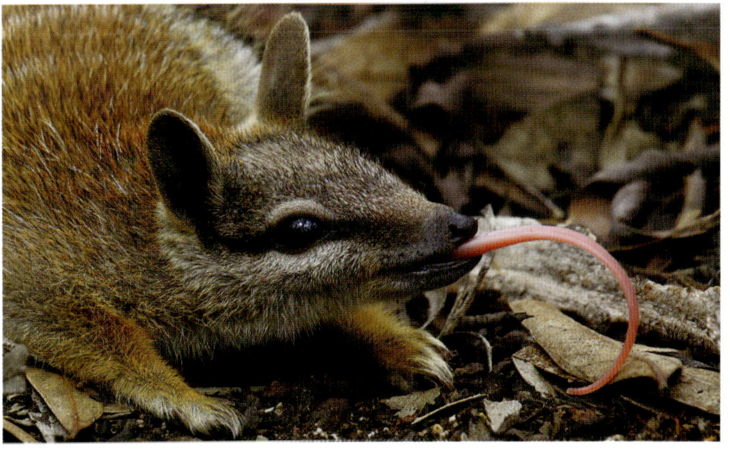

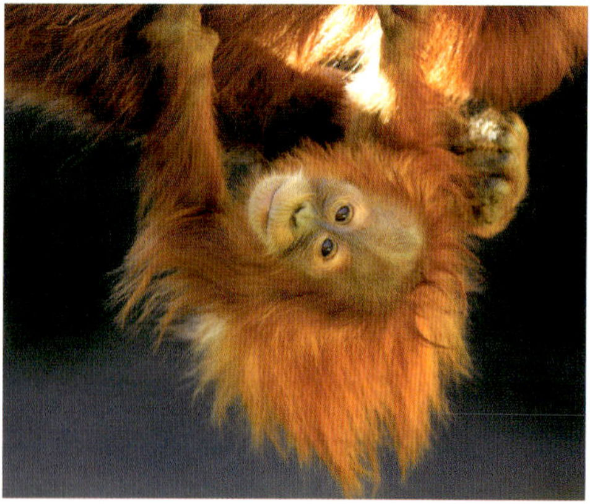

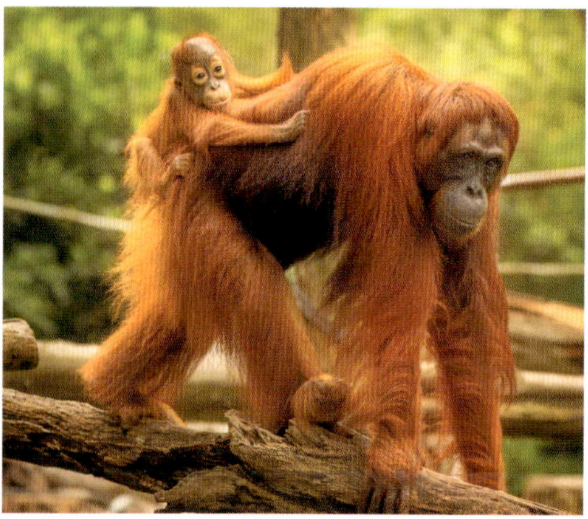

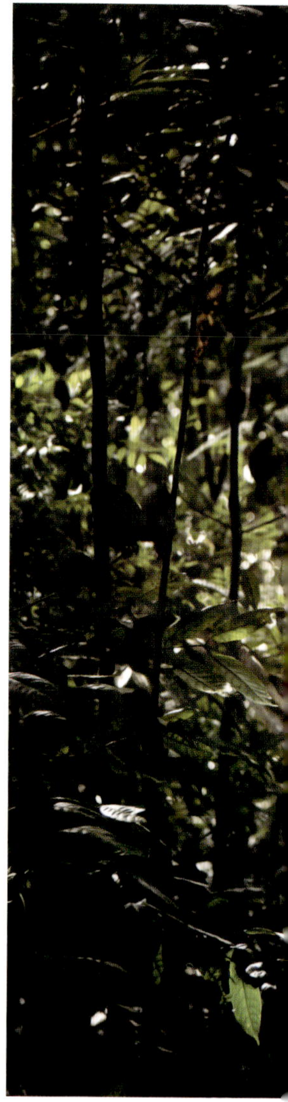

ALL PHOTOGRAPHS:

Orang-utan

The only great ape that survives in Asia (other than humans, of course), both species of orang-utan are endangered. As well as a shrinking habitat, the apes are threatened by hunters who kill the mother and take the young for sale as pets. Orang-utans breed very slowly, with babies born nine years apart, so it will take many decades for populations to recover.

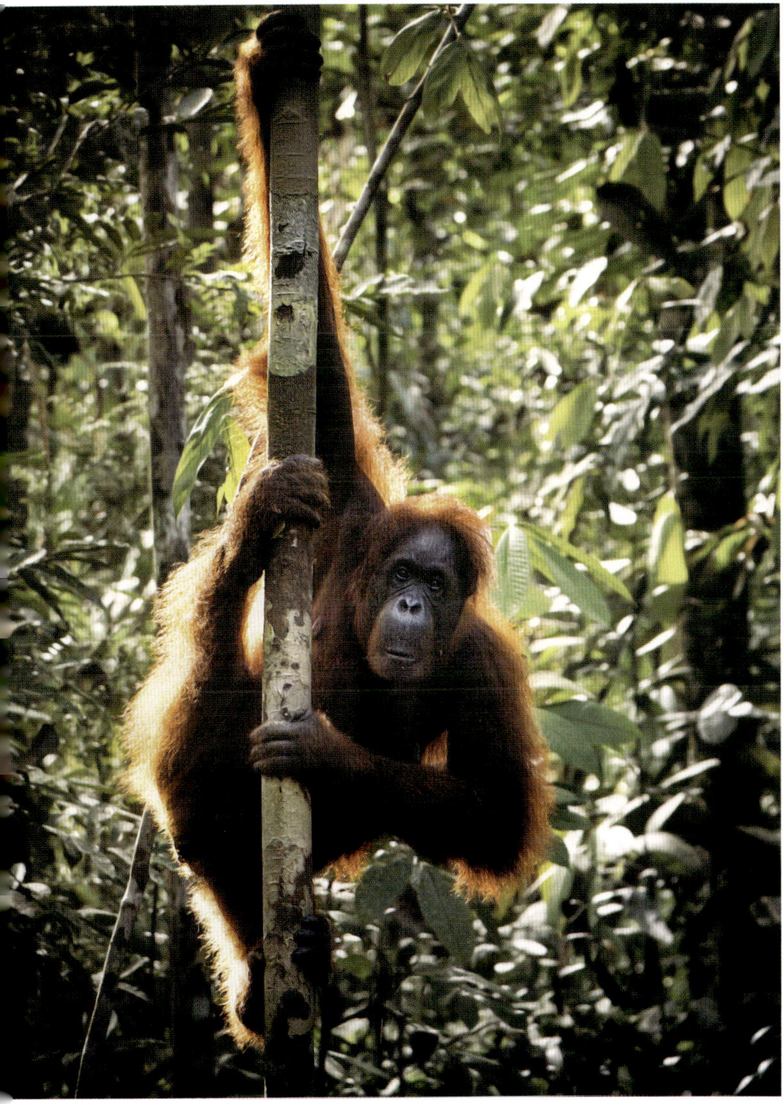

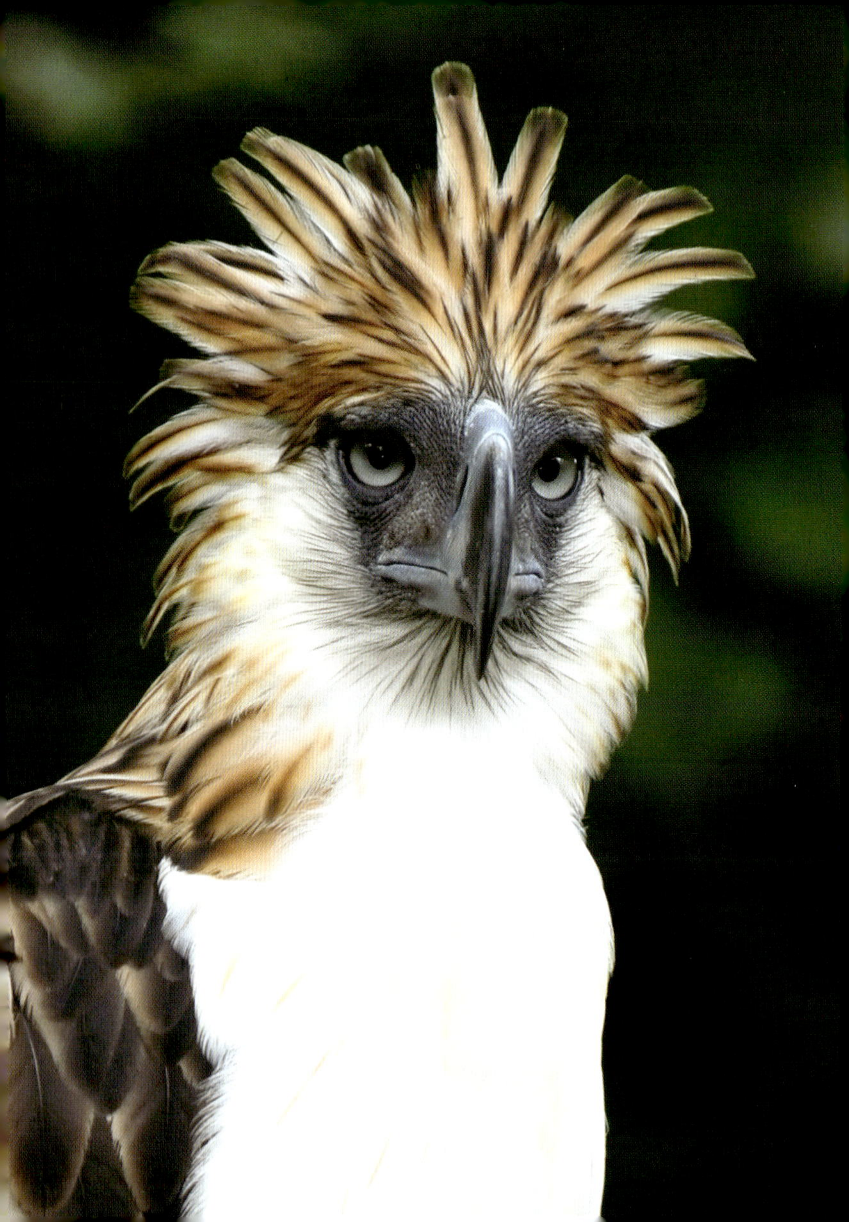

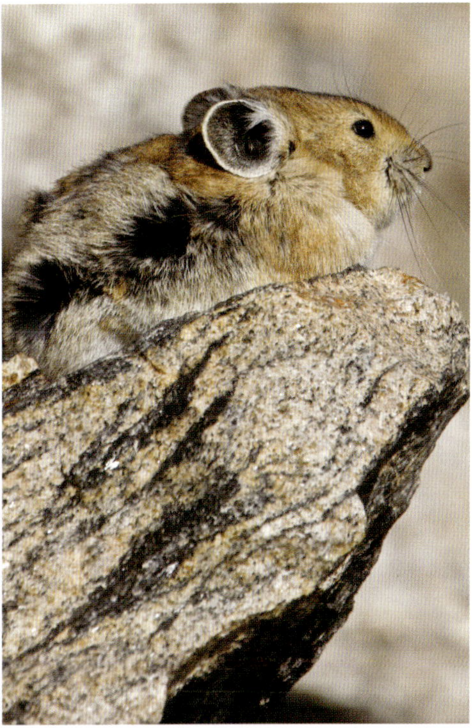

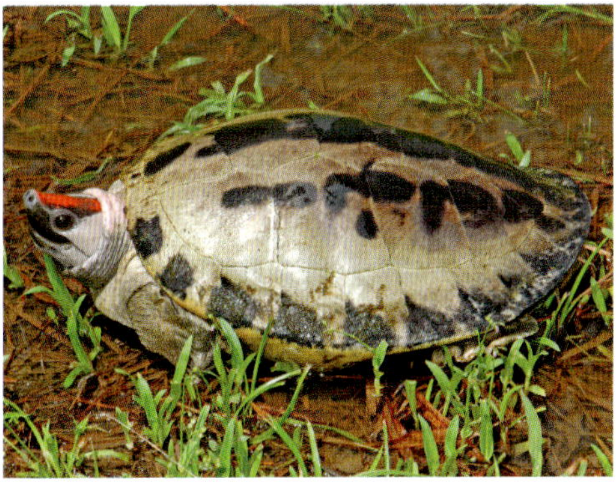

OPPOSITE:

Philippine eagle

The largest bird of prey of all, this forest hunter swoops in to snatch monkeys from the branches. They are threatened by extreme deforestation and are critically endangered. The wild population could be as low as 200.

LEFT:

Pika

These diminutive relatives of rabbits and hares live across Central Asia (and North America), often in rocky habitats. Many of them require protection.

BELOW:

Painted terrapin

Found across the forests of Indonesia and Southeast Asia, this freshwater turtle is threatened by the widespread collection of its eggs for food and for supplying the pet market. Today, its eggs are collected by rangers who ensure they hatch in captivity before returning them to the wild.

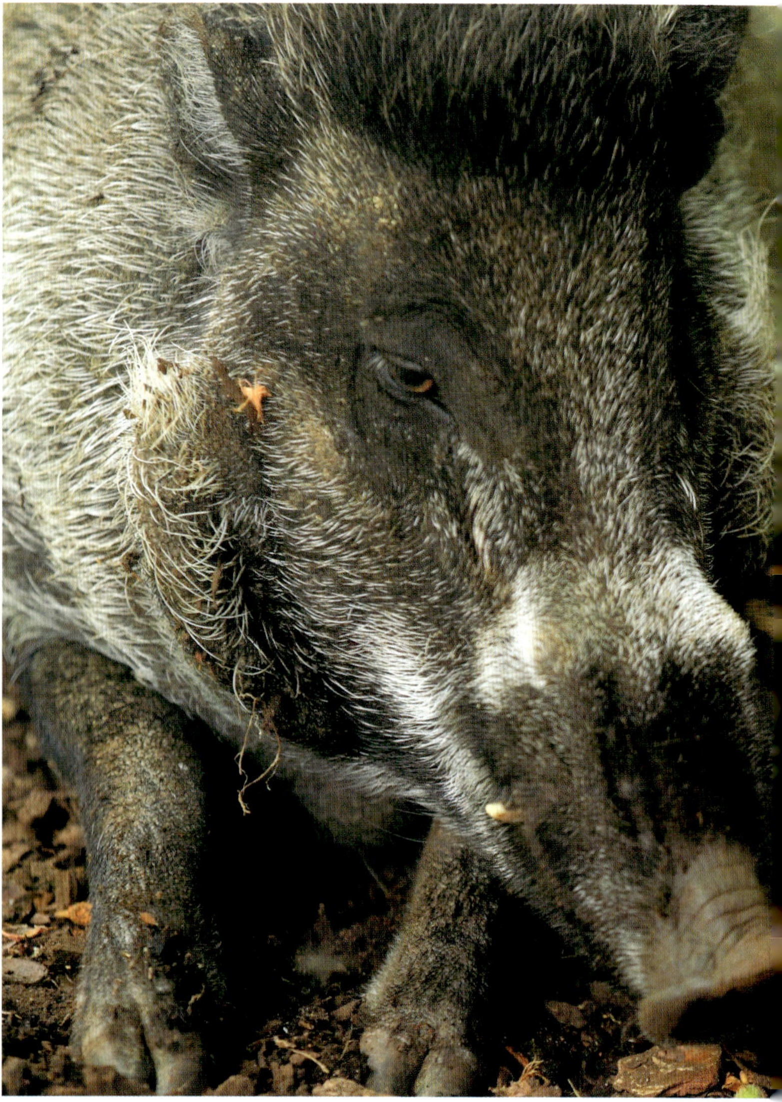

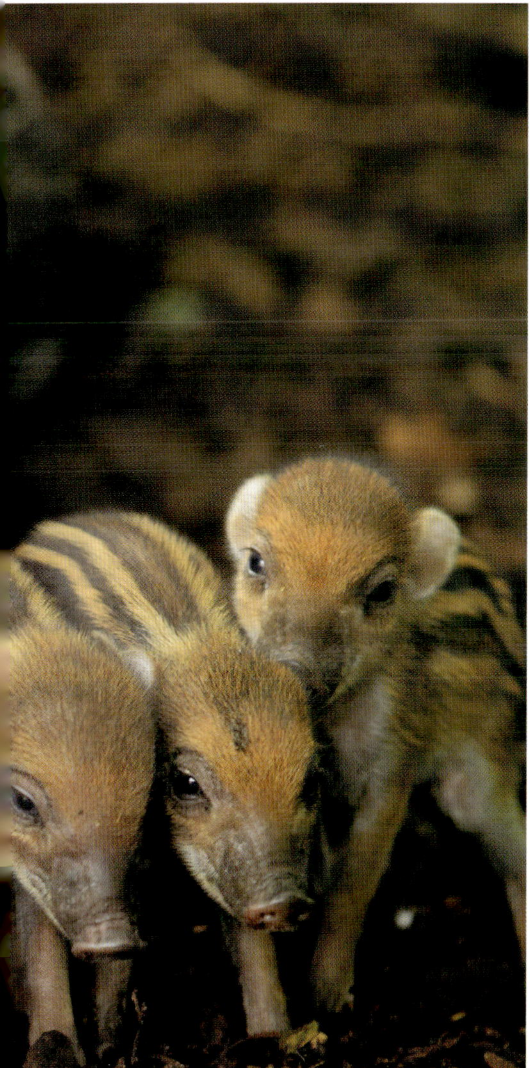

Visayan warty pig

This rare wild pig from the forests of a few islands in the Philippines is critically endangered. As its habitat has been systematically removed, the wild population of this pig has dwindled by 98 per cent. There are several captive breeding programmes running around the world.

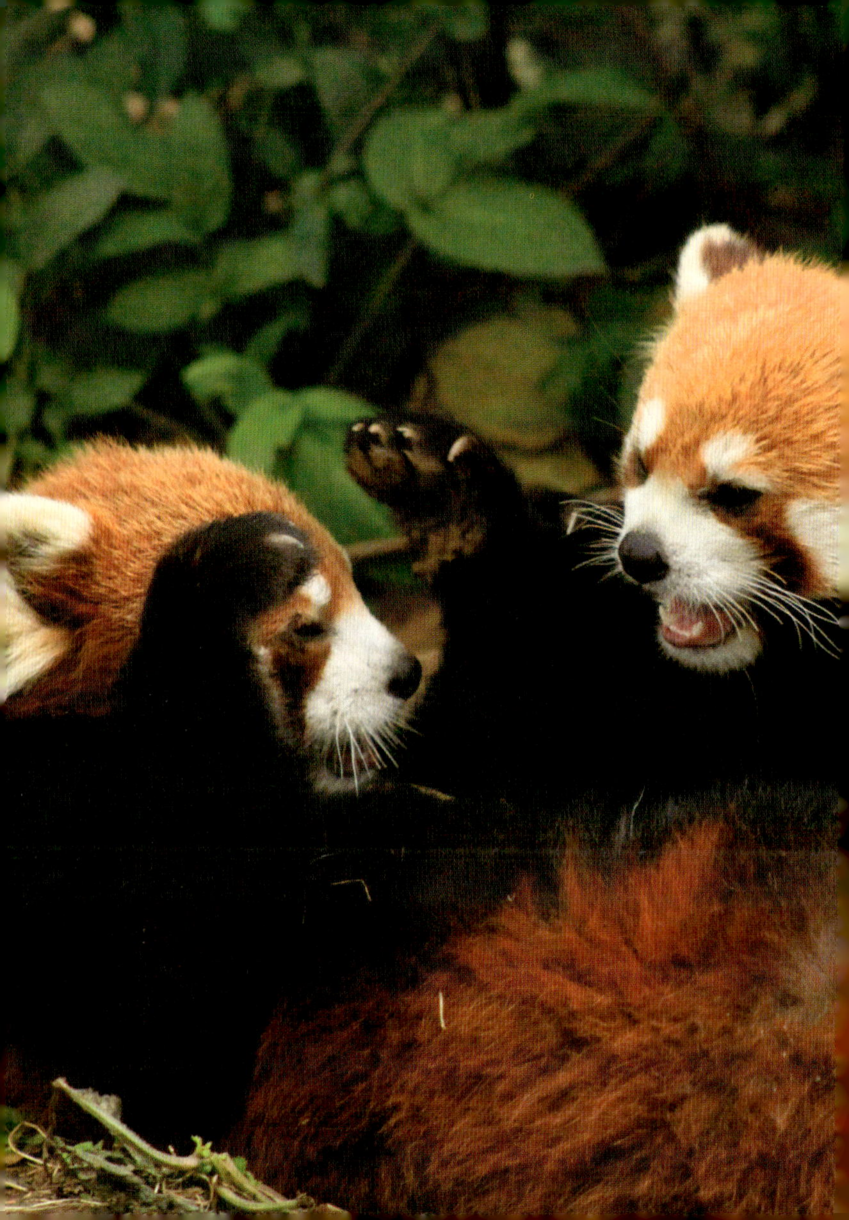

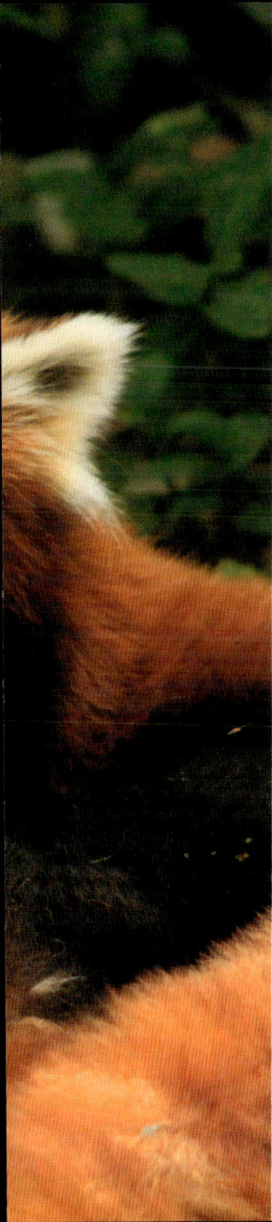

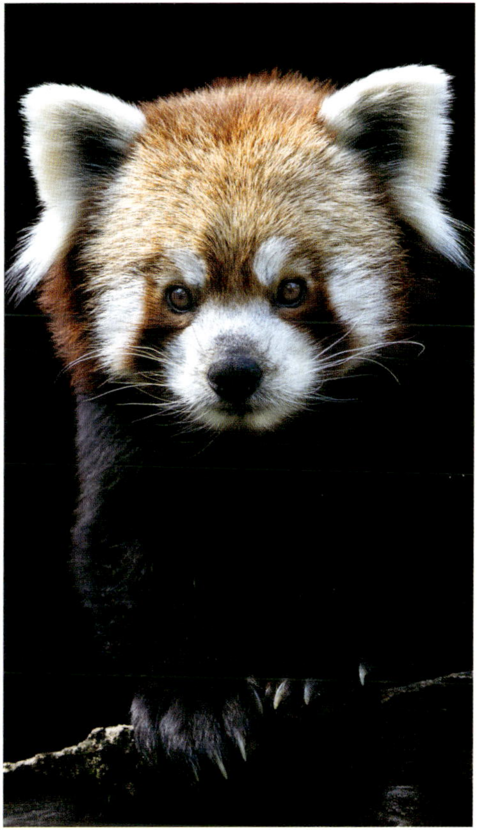

ALL PHOTOGRAPHS:

Red panda

Earning the panda name from the cute white patches on its face, this Chinese species is not related to the giant panda. Instead it is linked to the raccoons of the Americas. Like the giant panda, the red panda relies on a similar bamboo diet. The wild population is below 10,000 due to habitat loss and hunters wanting its thick red fur.

Siamese crocodile
Very few Siamese crocodiles survive in the wild, although there are many thousands in captivity. They have been driven out of most of their range and survive in a few remote mountain areas.

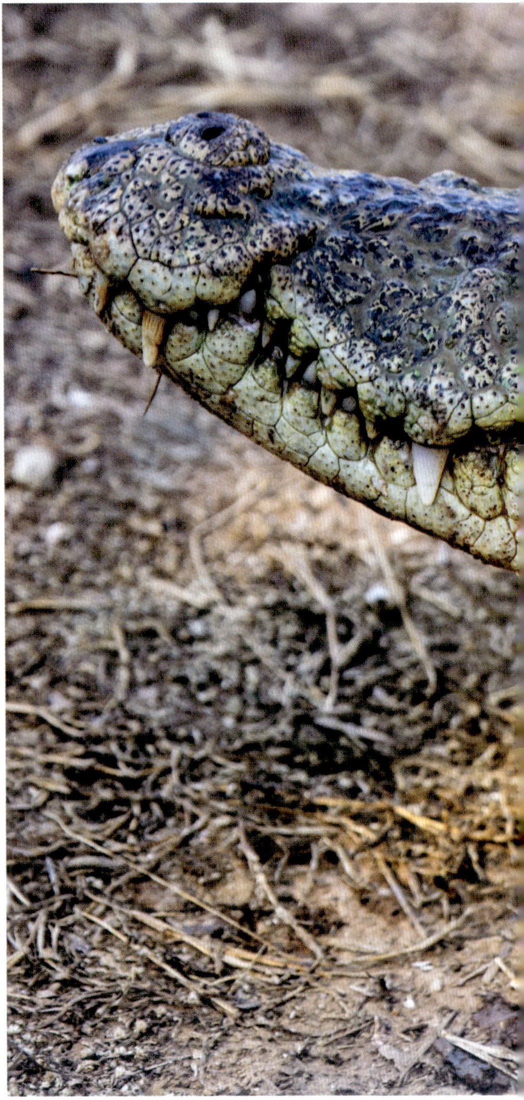

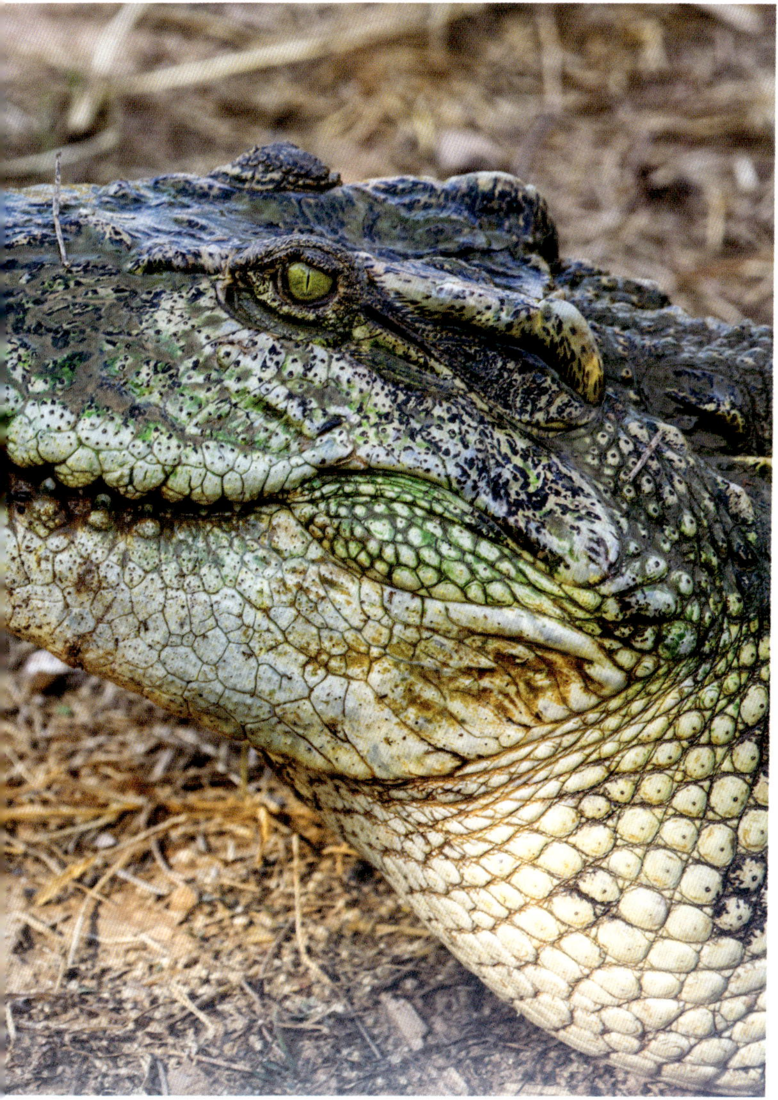

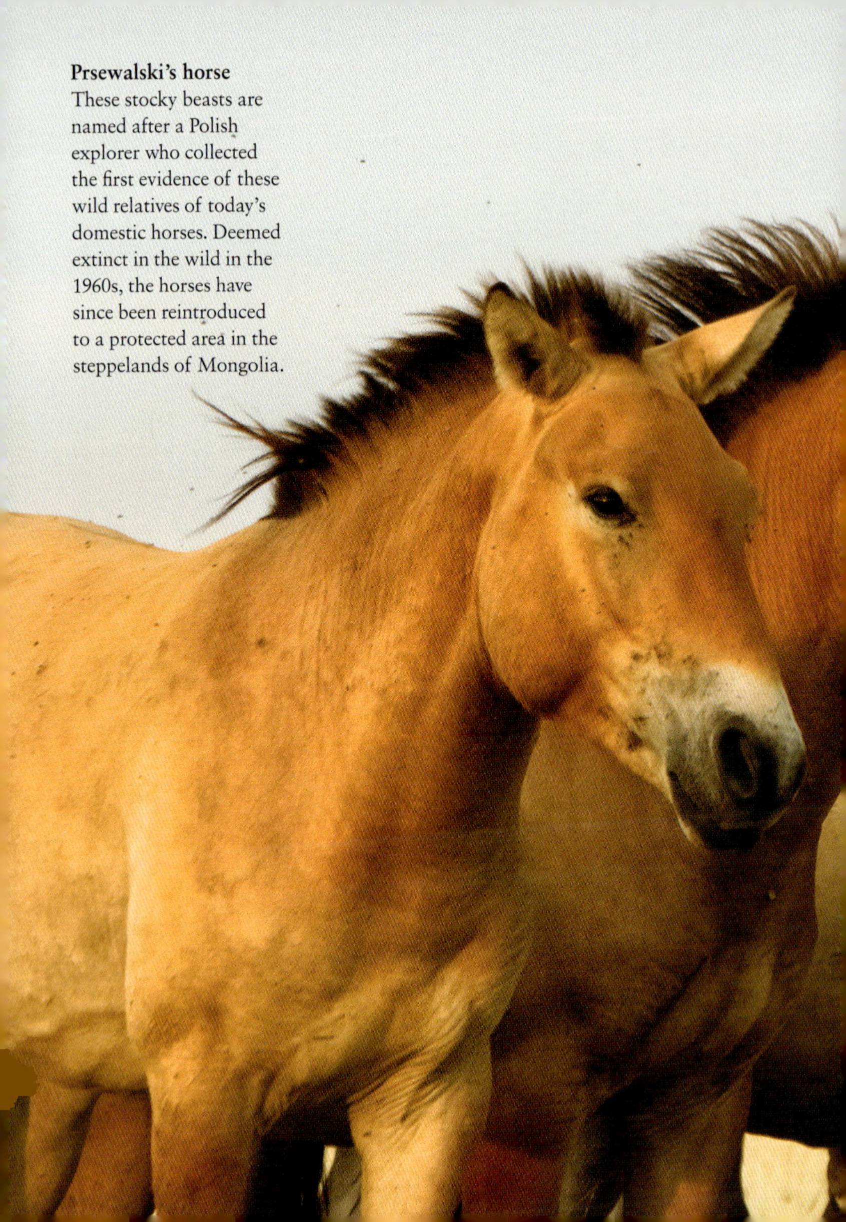

Prsewalski's horse
These stocky beasts are named after a Polish explorer who collected the first evidence of these wild relatives of today's domestic horses. Deemed extinct in the wild in the 1960s, the horses have since been reintroduced to a protected area in the steppelands of Mongolia.

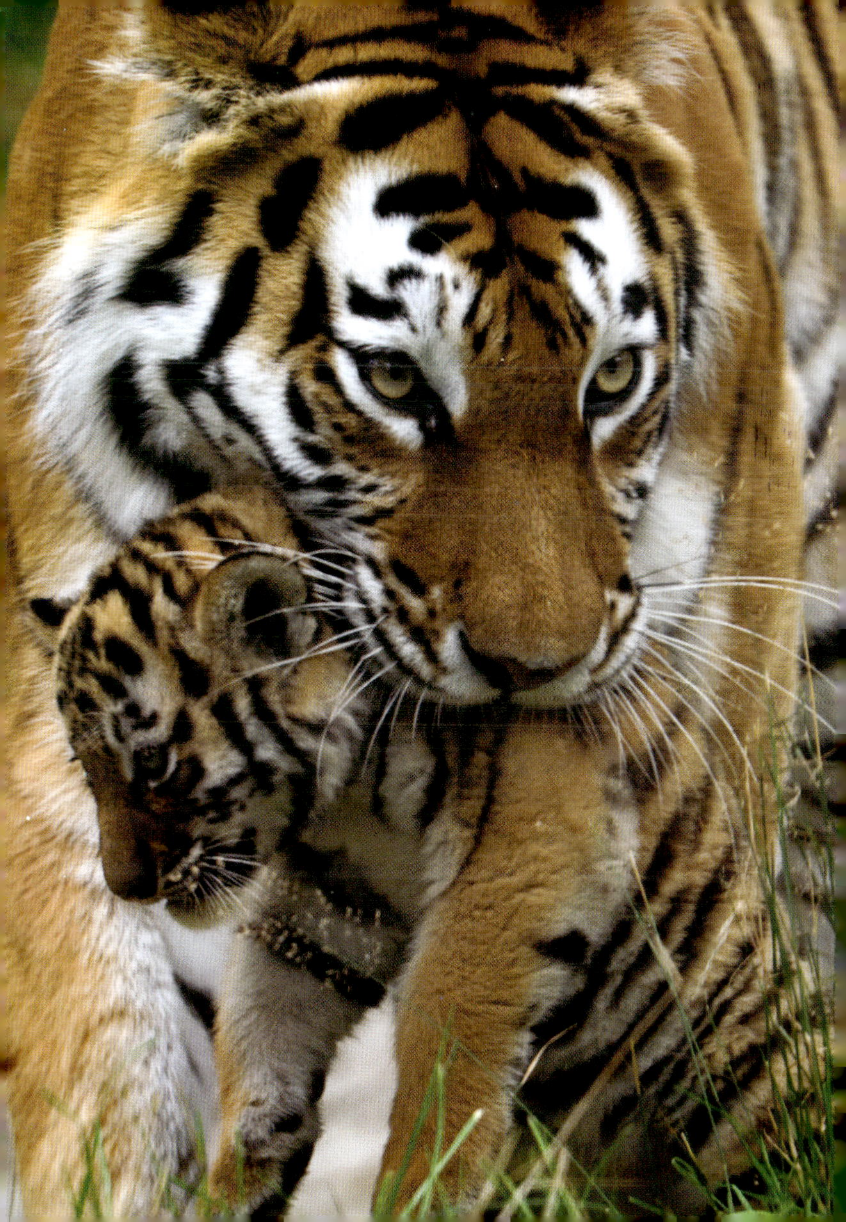

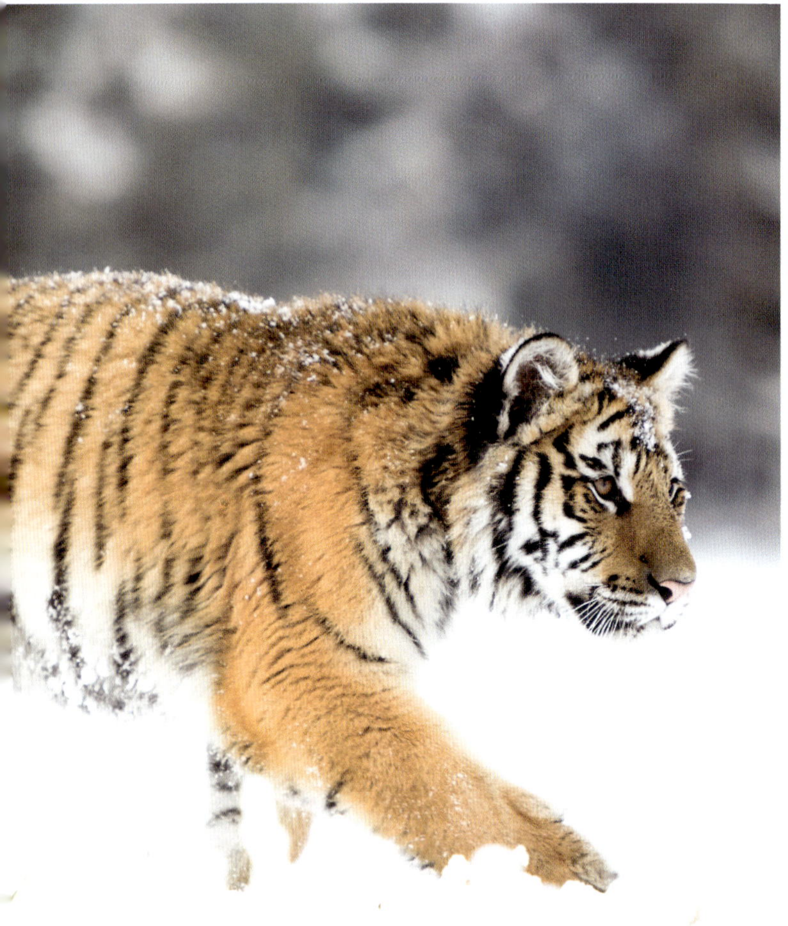

Siberian tiger
Among the largest of the tiger subspecies, the Siberian tiger prowls the cold forests of Manchuria in the far north of China, and Siberia in Russia. In recent years its population has increased to about 500 breeding adults.

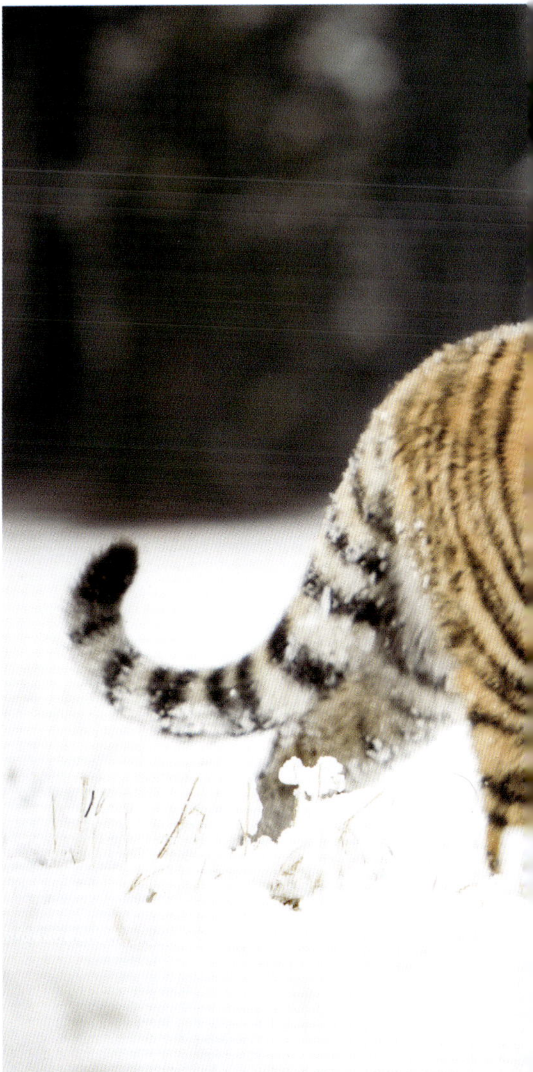

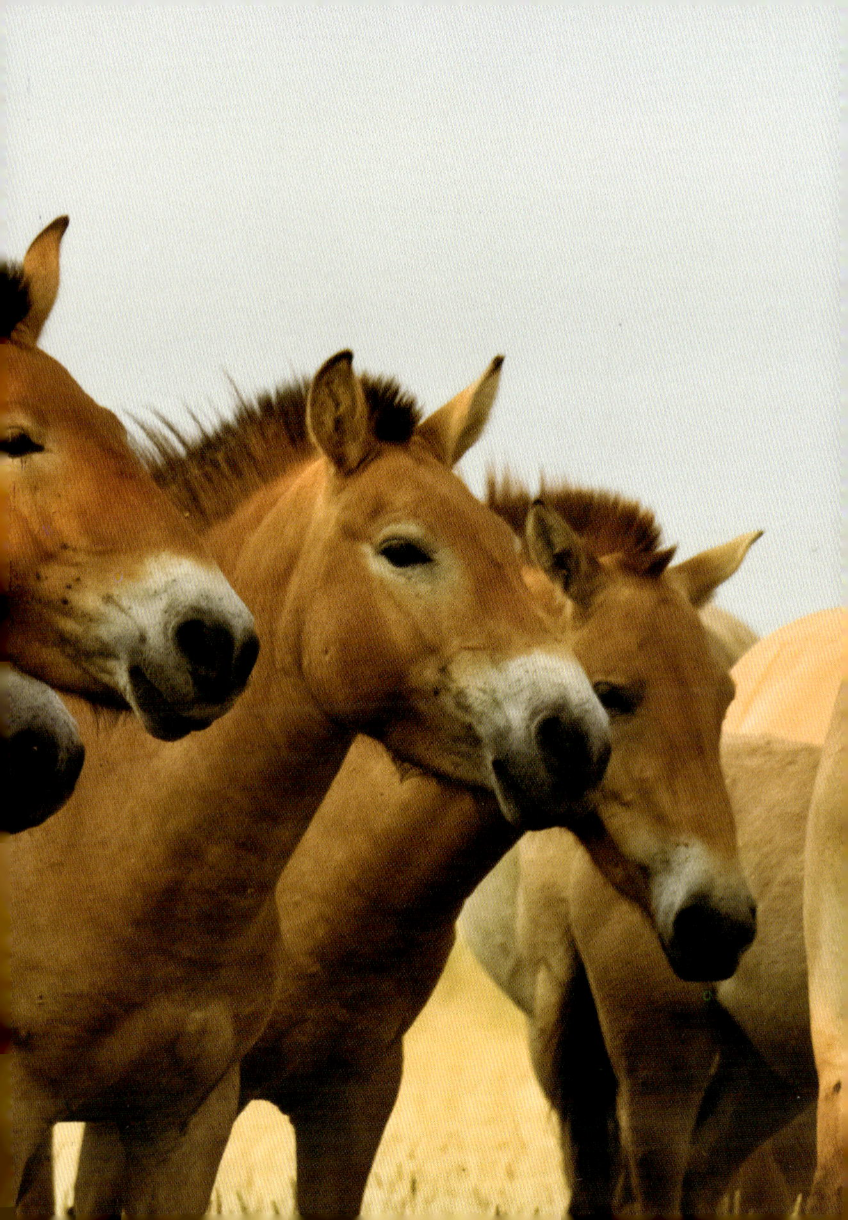

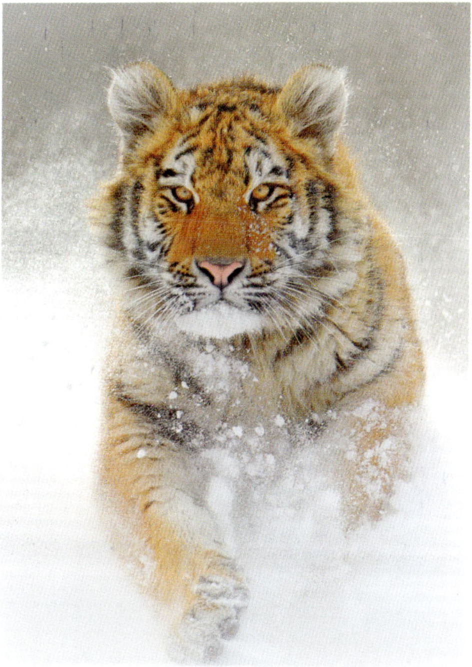

ALL PHOTOGRAPHS:
Siberian tiger
These tigers are highly protected in Russia, where most of them live today. The numbers in China and North Korea are less clear but thought to be very low.

OVERLEAF:
Pangolin
Looking more like a reptile than a mammal thanks to the scales that armour its body, the pangolins of Asia are all endangered – as are their relatives in Africa. The chief threat is collection from the wild for use in spurious traditional East Asian medicines.

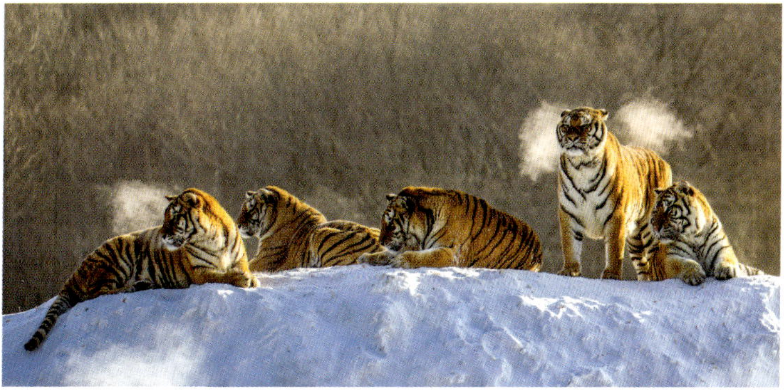

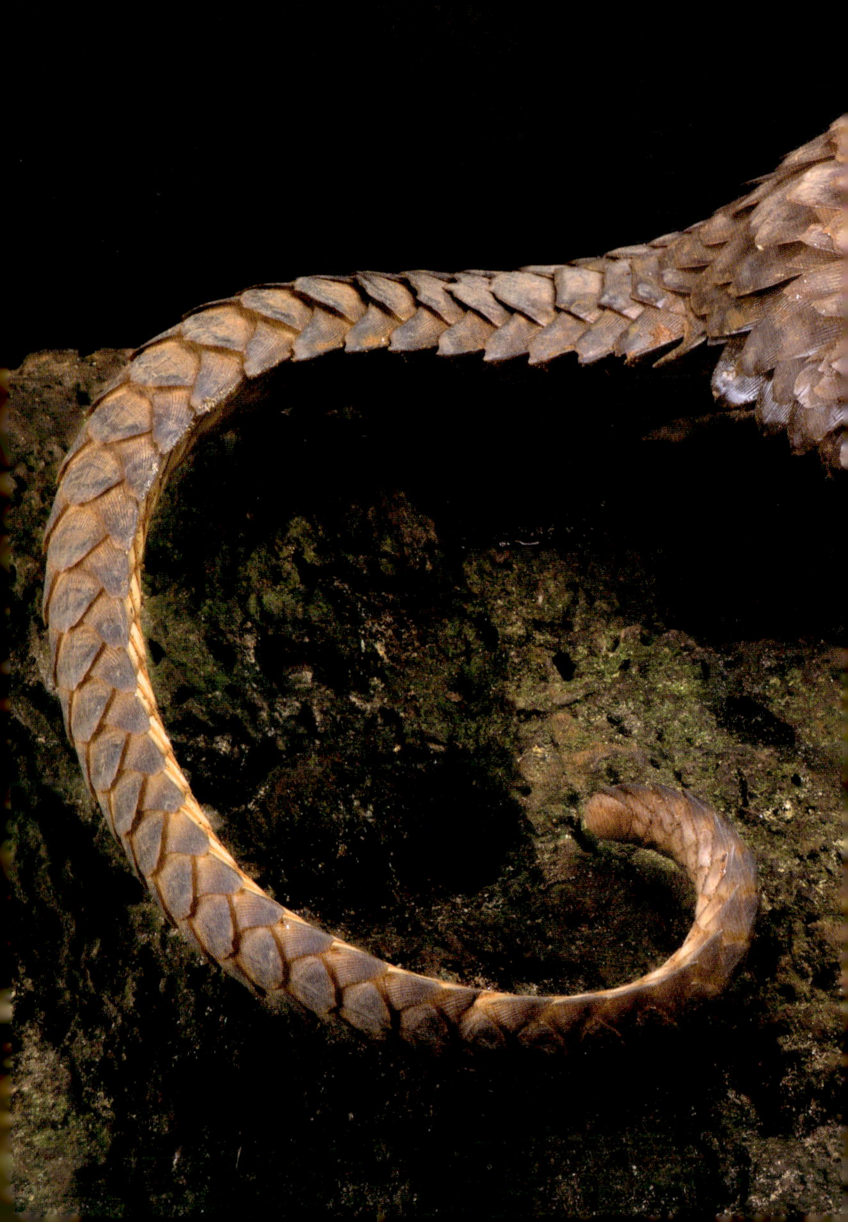

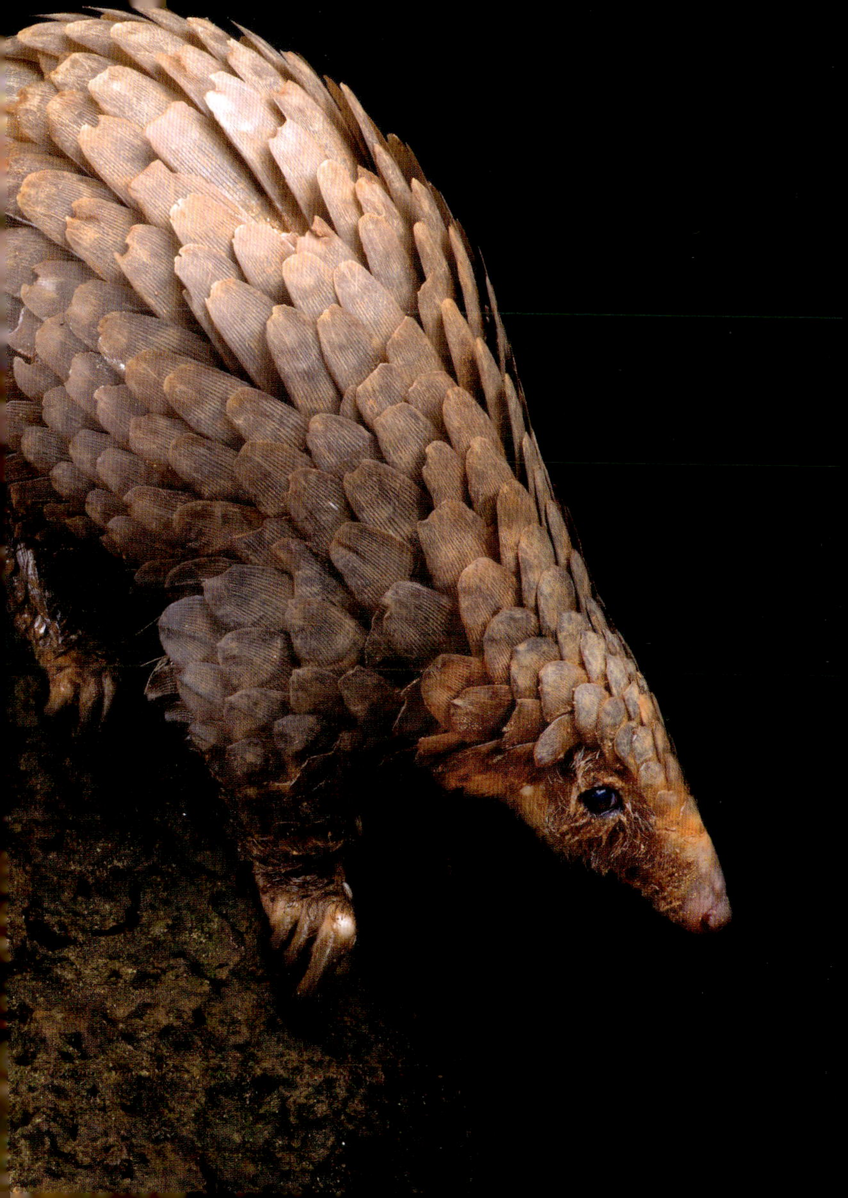

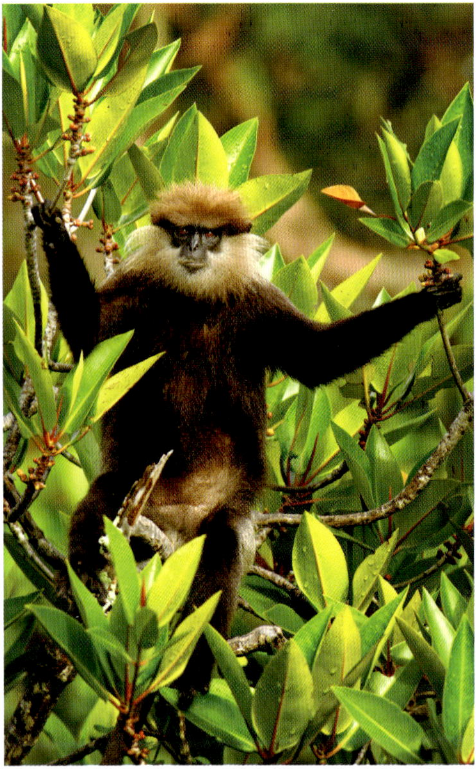

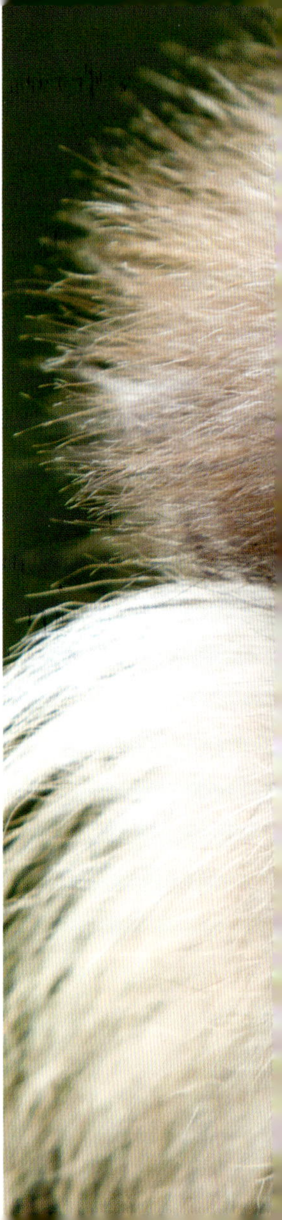

ALL PHOTOGRAPHS:

Purple-faced langur

This endangered monkey endemic to Sri Lanka is threatened by the fragmentation of its habitat. In recent years, forest corridors have been maintained so the monkeys can continue to spread out through a wide area.

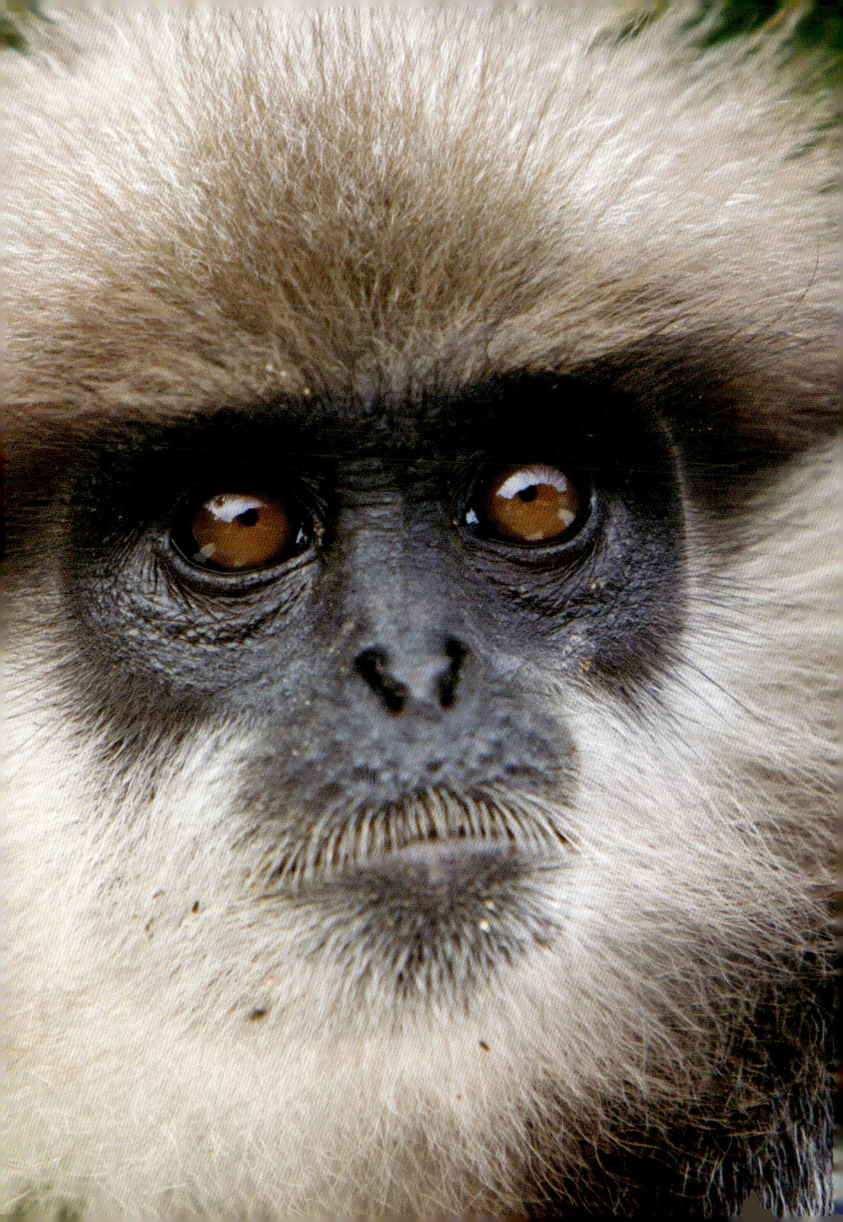

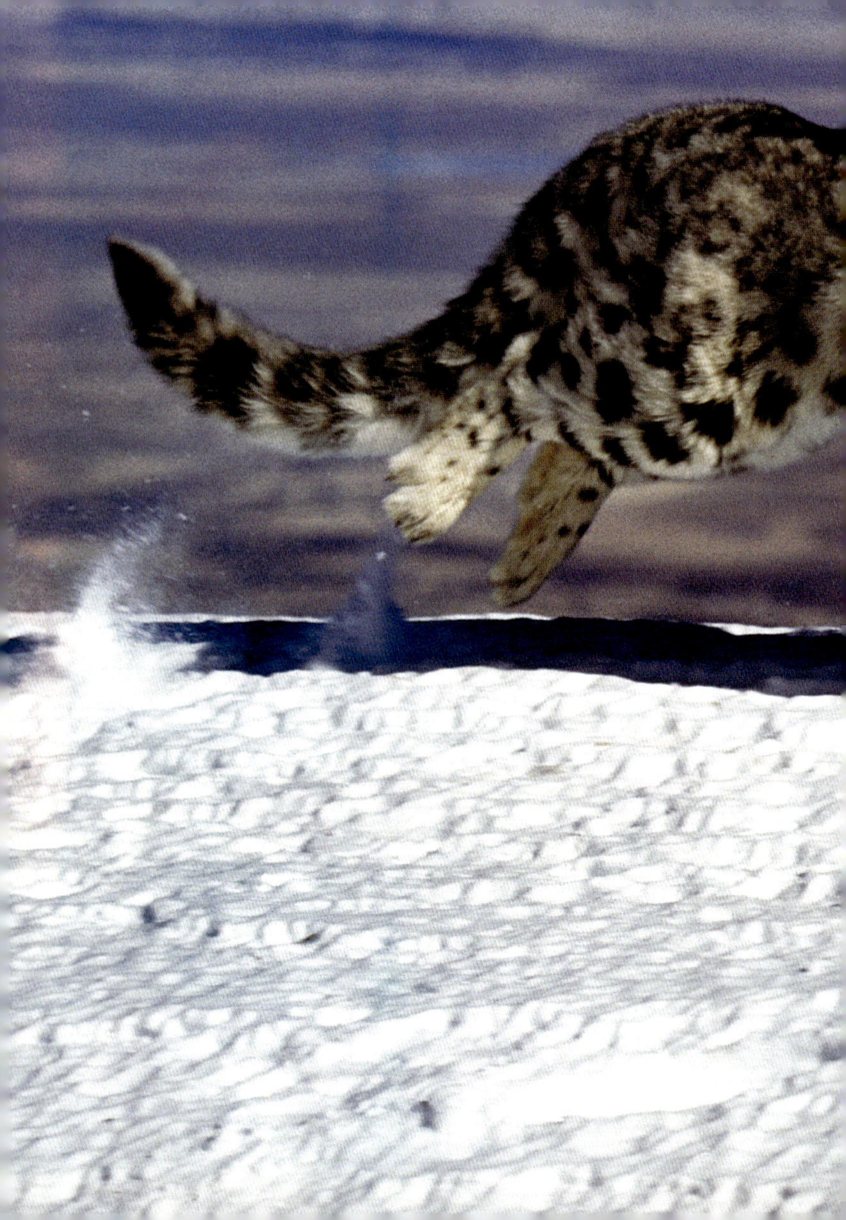

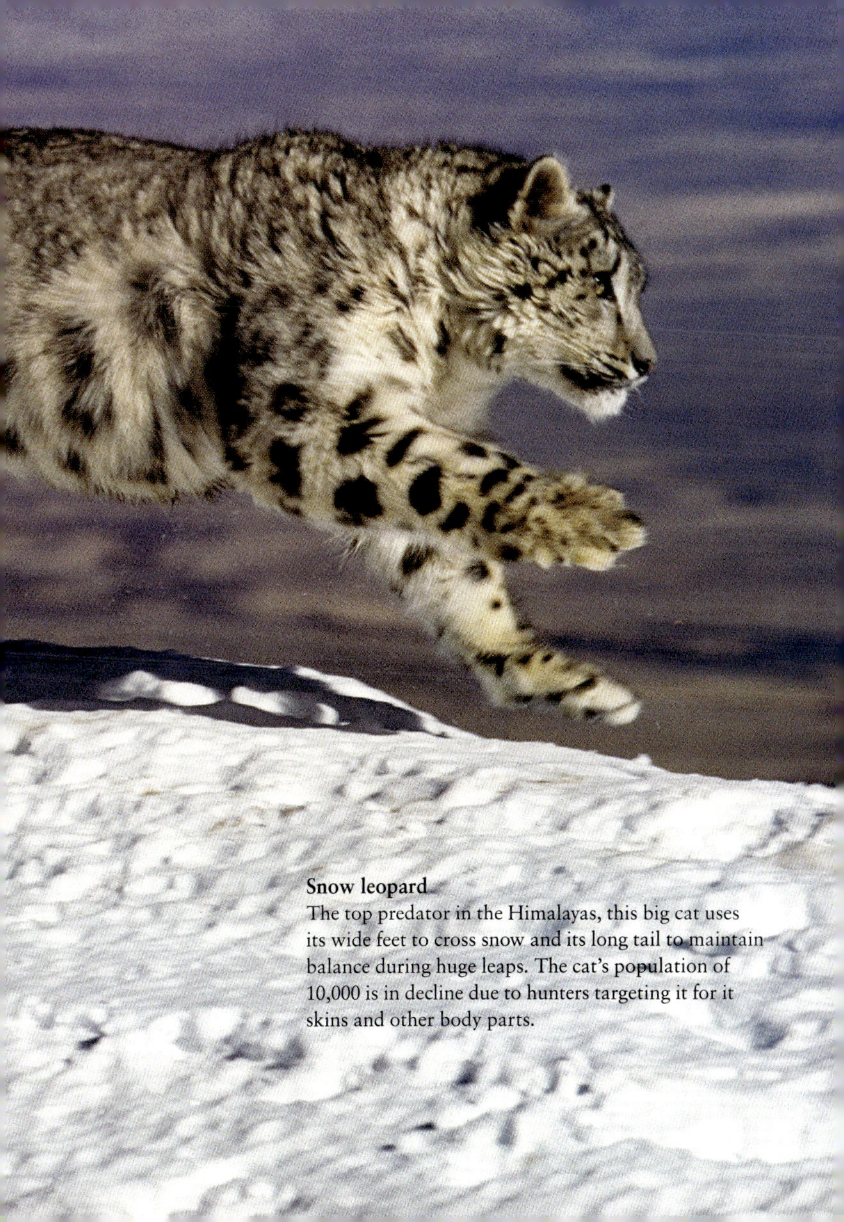

Snow leopard
The top predator in the Himalayas, this big cat uses its wide feet to cross snow and its long tail to maintain balance during huge leaps. The cat's population of 10,000 is in decline due to hunters targeting it for it skins and other body parts.

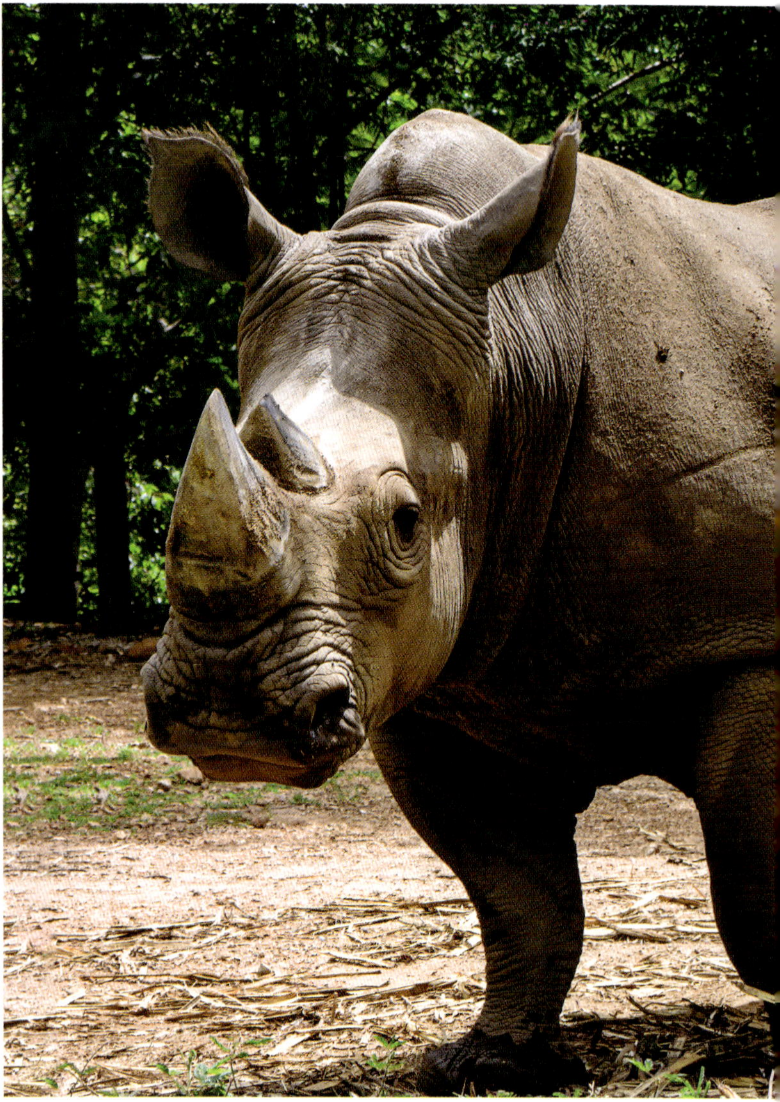

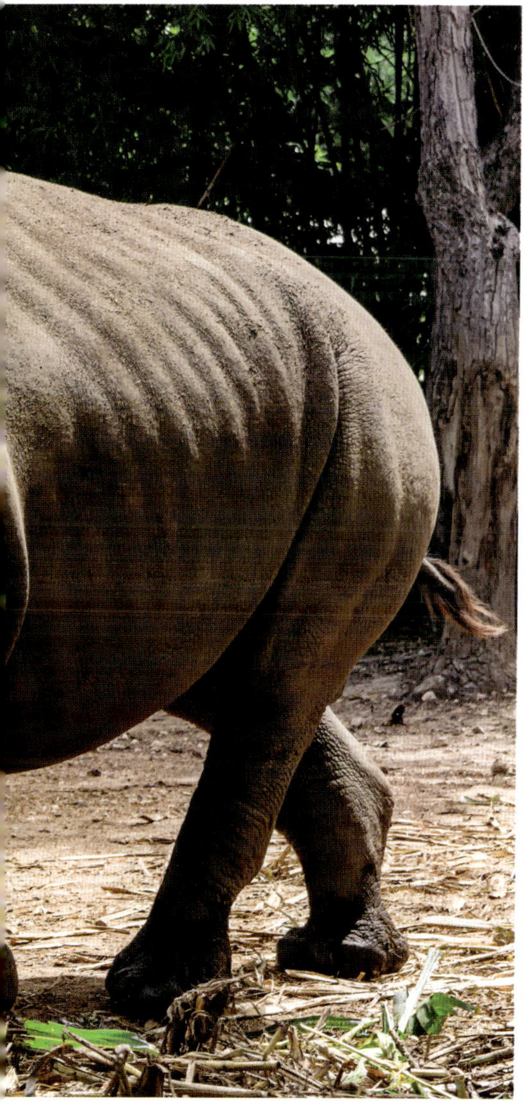

Sumatran rhinoceros
Only about 80 of these double-horned rhinos survive in the wild as their forest habitat is steadily destroyed.

Sumatran rhinoceros
Historically, hunters have targeted these animals for their horns for use in traditional medicines. It is also possible that the legend of the unicorn is derived from this and other Asian rhinos.

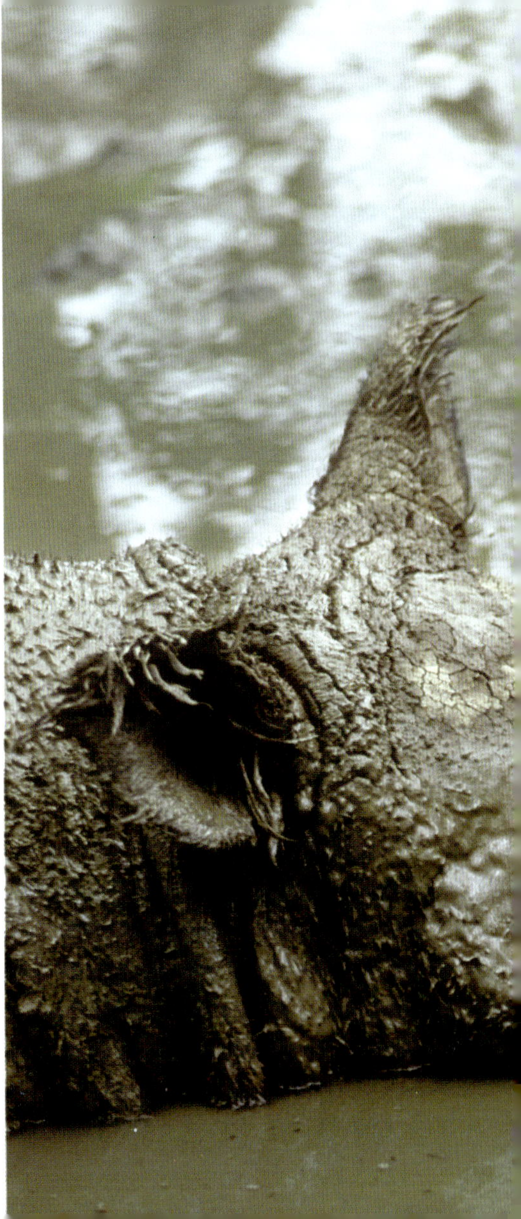

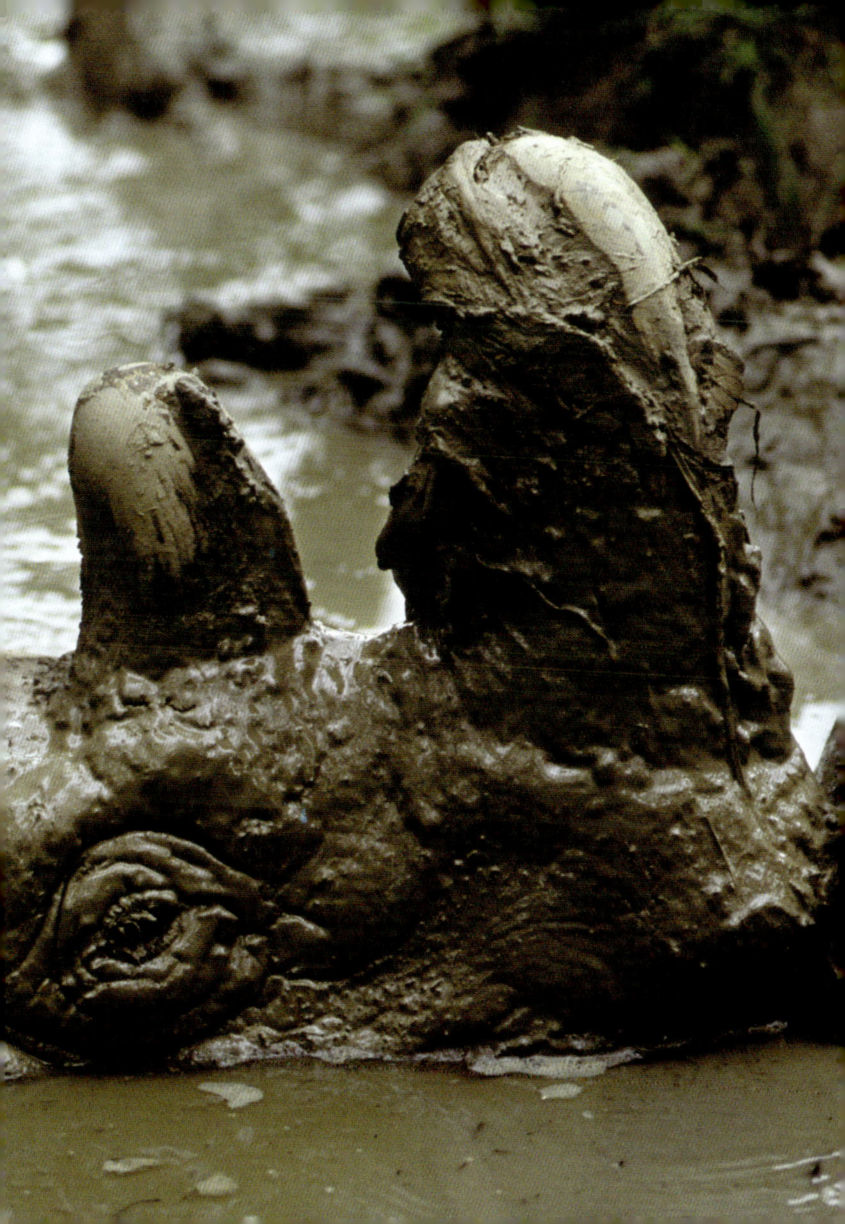

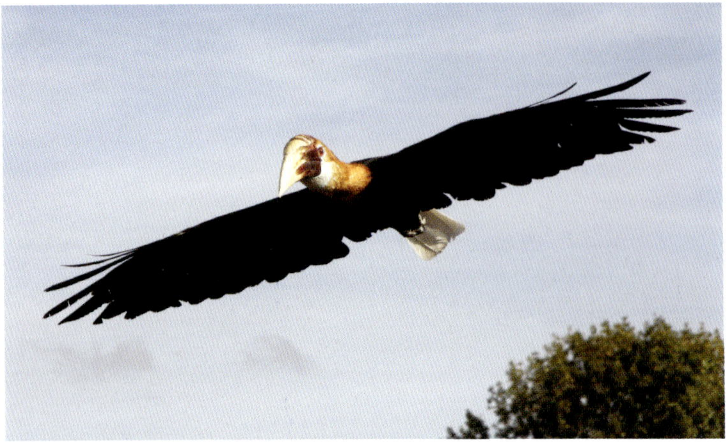

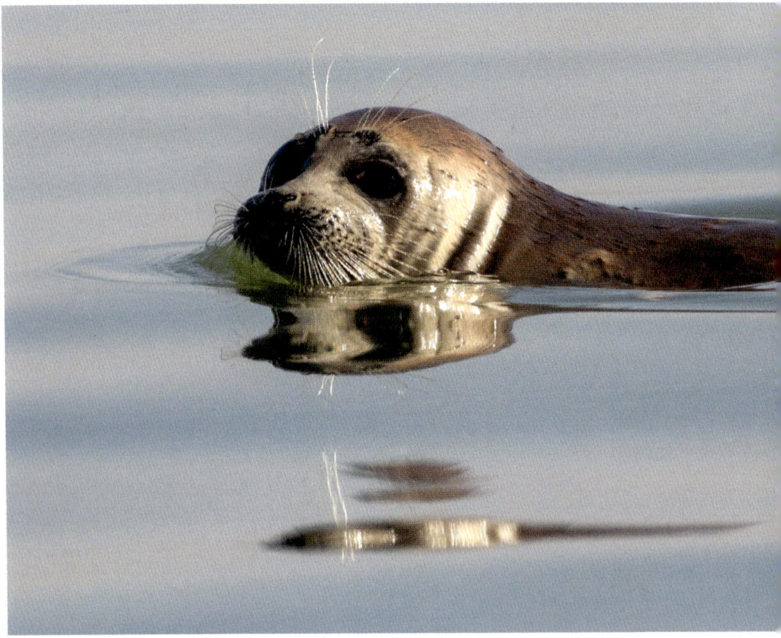

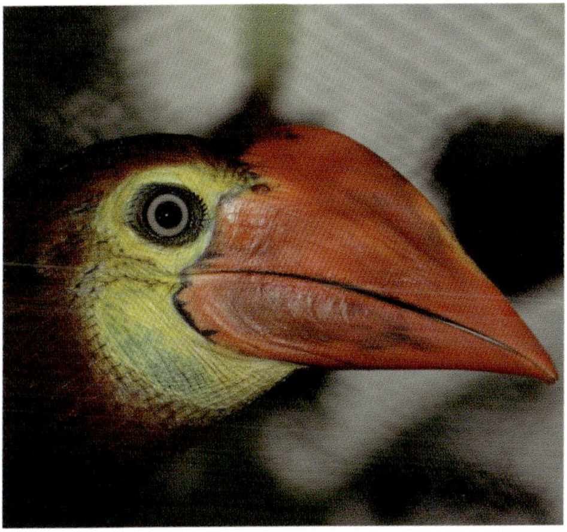

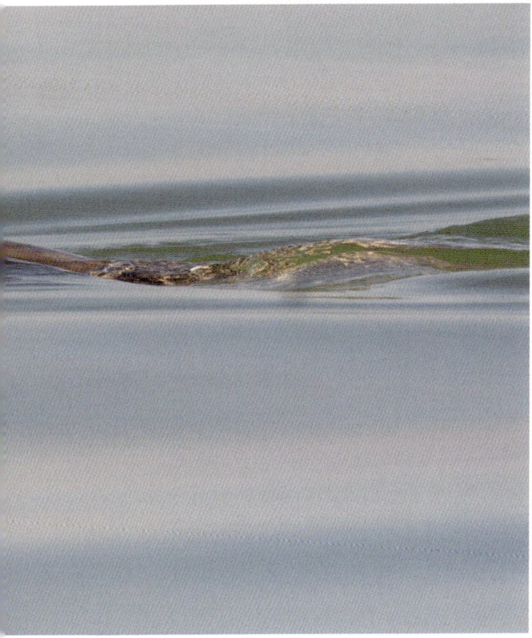

LEFT TOP AND OPPOSITE:

Rufous-headed hornbill

Found only on two islands in the Philippines, their population is about 2,000 and still falling due to logging of its habitat.

LEFT BOTTOM:

Caspian seal

This unique species from the inland Caspian Sea is the only seal to live in a freshwater habitat. The population has been in steady decline as the water has become more polluted.

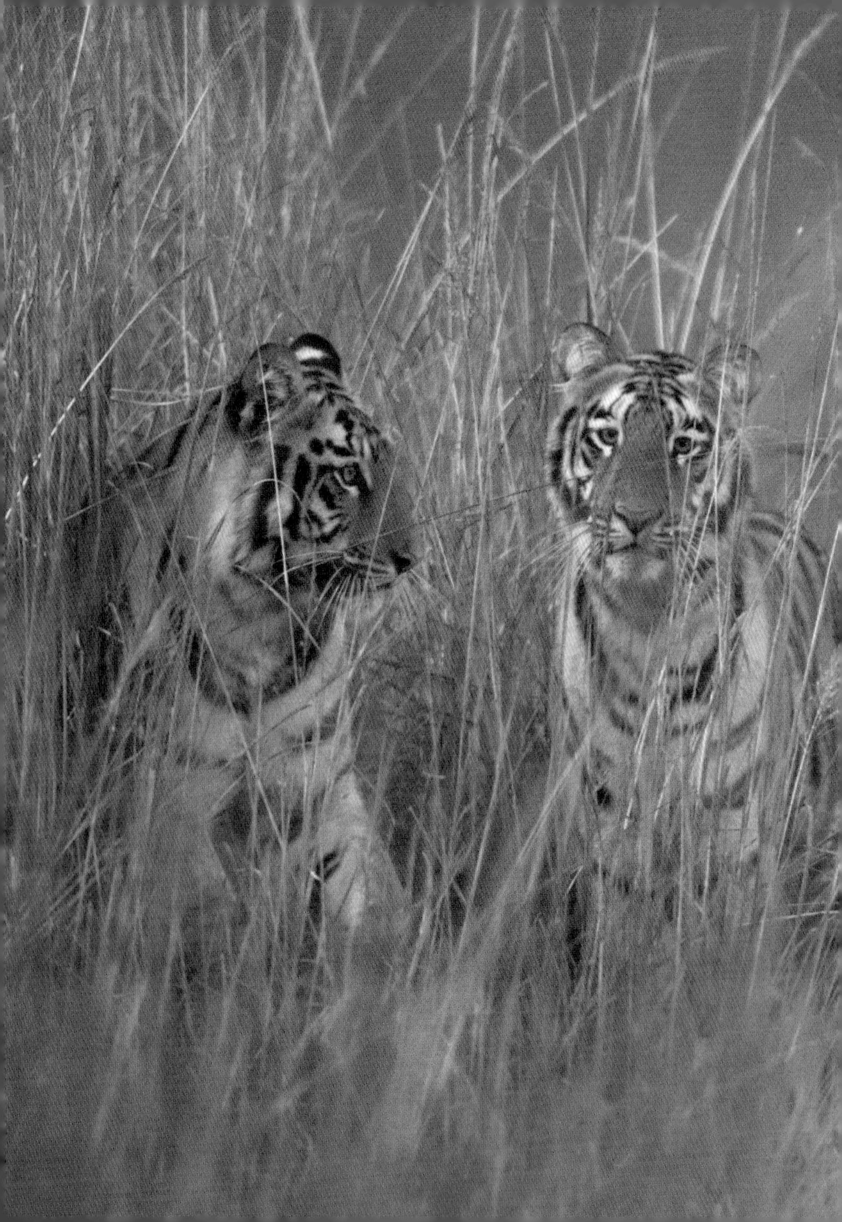

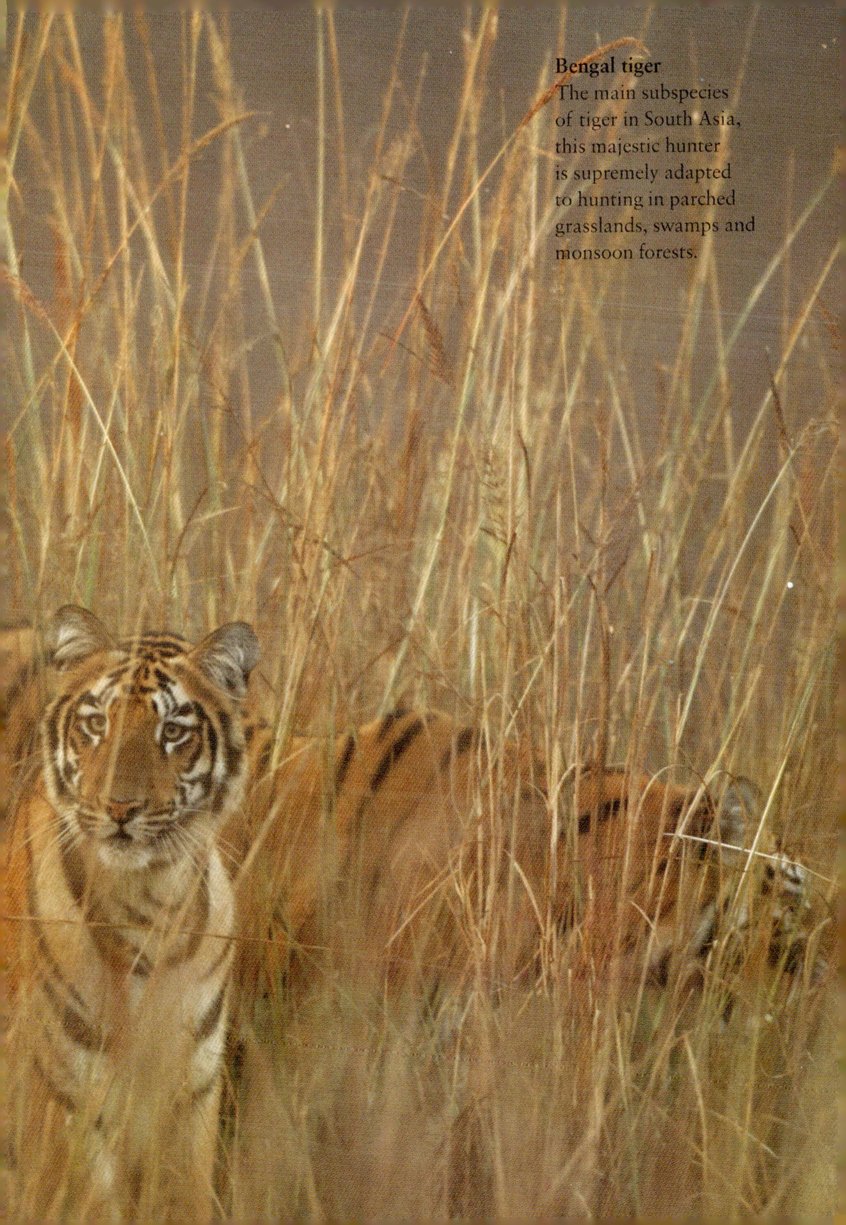

Bengal tiger
The main subspecies of tiger in South Asia, this majestic hunter is supremely adapted to hunting in parched grasslands, swamps and monsoon forests.

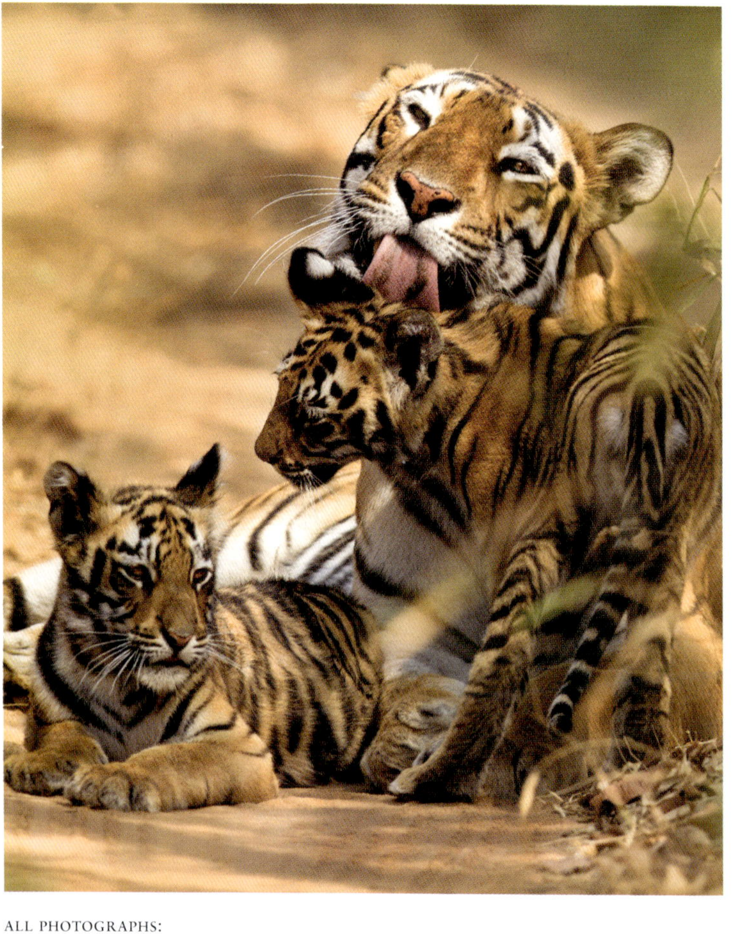

ALL PHOTOGRAPHS:

Bengal tiger

Poaching and habitat fragmentation means that the wild population of this mighty hunter has plummeted to less than 4,000. In recent years, protection programmes have been stepped up but the future of the most iconic of tigers still lies in the balance.

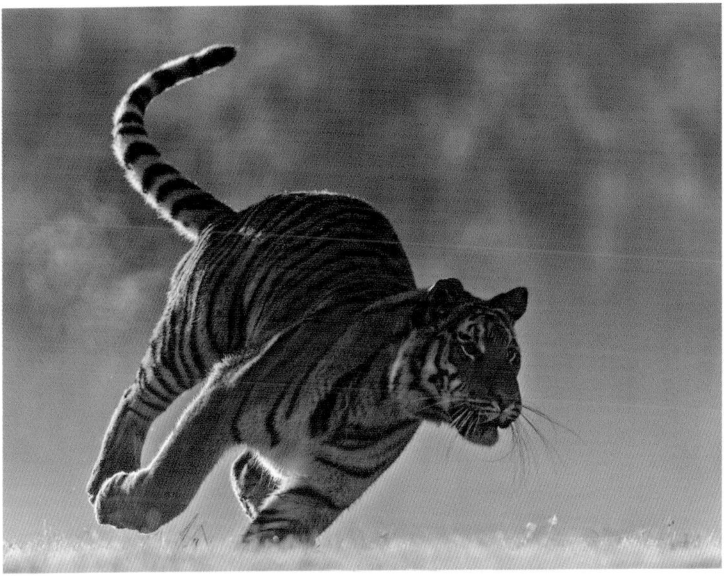

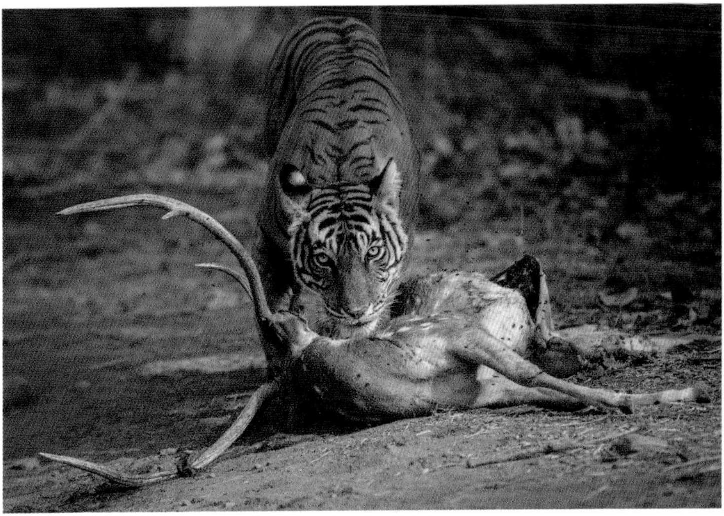

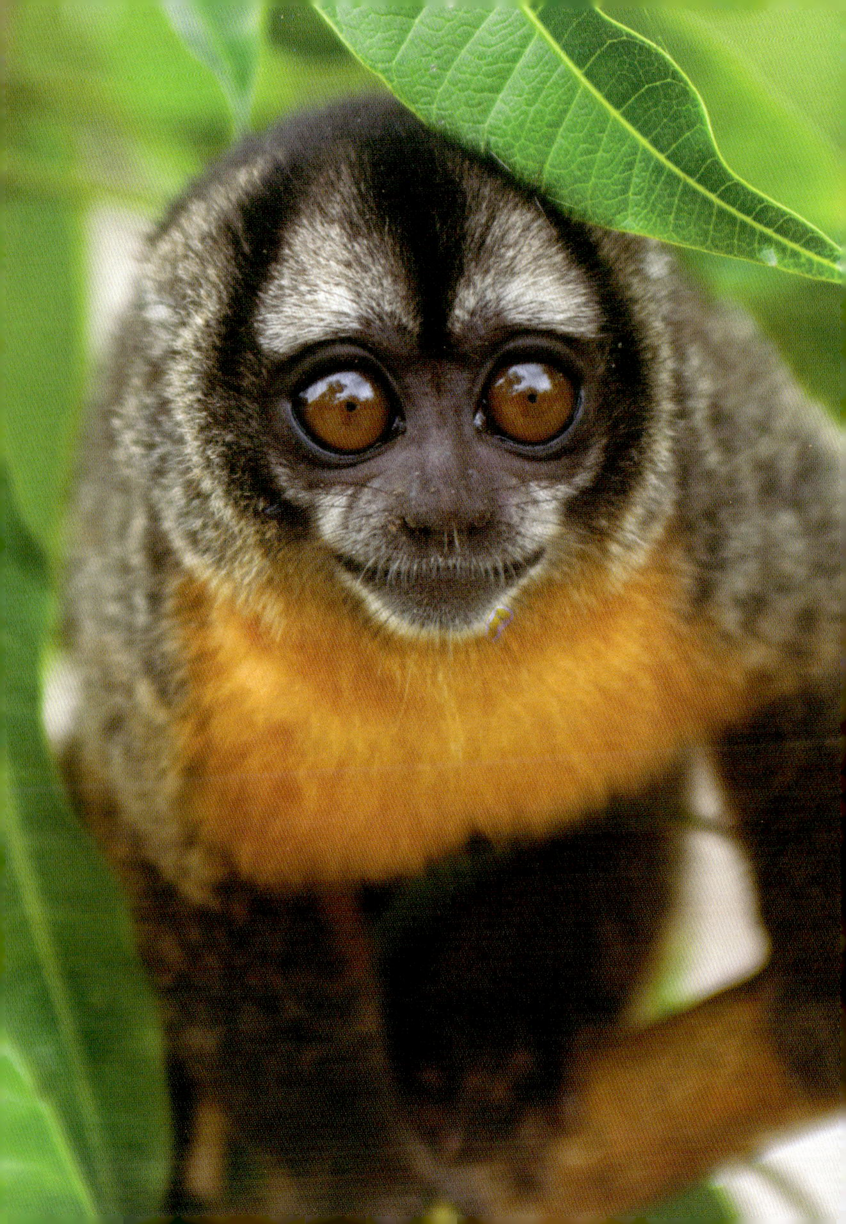

The Americas

North and South America are two entirely distinct continents that connect at the Isthmus of Panama. Both separated from the rest of the continents about 100 million years ago, but they remained apart from each other until only a few million years ago. As a result, the animals of North America are a closer match to the fauna of Europe and Asia, while down in South America, the animals are very different, closer to those in Australia. So there are marsupials in the jungles of South America! The wildlife of both continents seldom gets the chance to migrate to its neighbouring territory due to a dense patch of mountainous jungle in Panama that bars its way, and so their unique character persists.

North America has a longer history of habitat degradation, and many of the animals that lived wild in its vast prairies and plains are now at risk as this land has been largely turned over to agriculture. However, this ecological damage is dwarfed by the more recent activities in South America, most notably in the Amazon. The Amazon Basin supports the world's largest rainforest, which is home to an estimated third of all Earth's species. The ancient forest contains so much life that the animals living in one patch of forest are entirely different to those only a few kilometres away. Tragically, these priceless ecological wonders are being destroyed on an industrial scale, and few Amazonian species are safe from extinction.

OPPOSITE:
Night monkey
The Americas are home to the only monkeys that come out at night when all other primates are settling down to sleep. This species from Peru is especially at risk as its forest habitat shrinks.

ALL PHOTOGRAPHS:

Amazon river dolphin

Also known as the boto, the famous pink dolphin of the Amazon is highly endangered, and its numbers continue to drop as it is killed in fishing nets. In recent years a new threat has emerged as local people kill the dolphin deliberately to act as bait for fish. Numbers are unknown but it is estimated that the population has now plunged to a quarter of its 2000 levels.

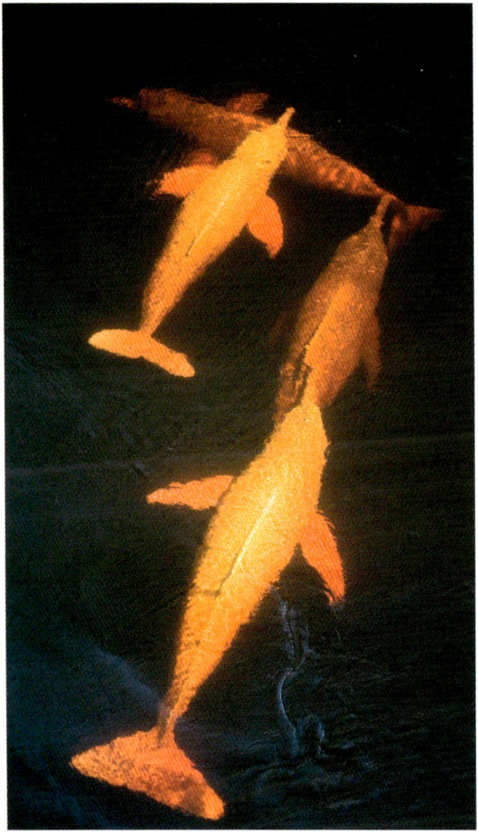

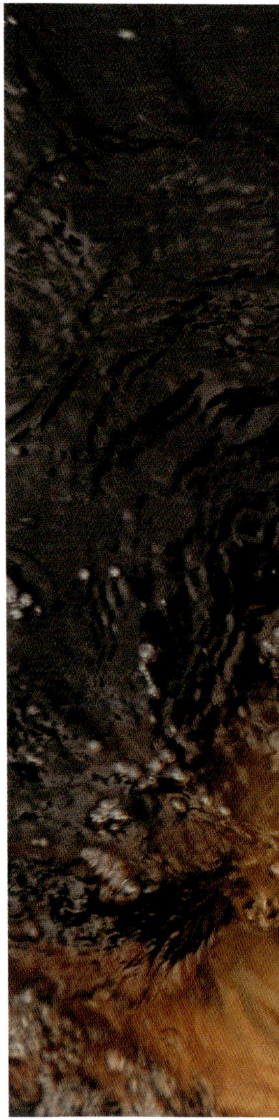

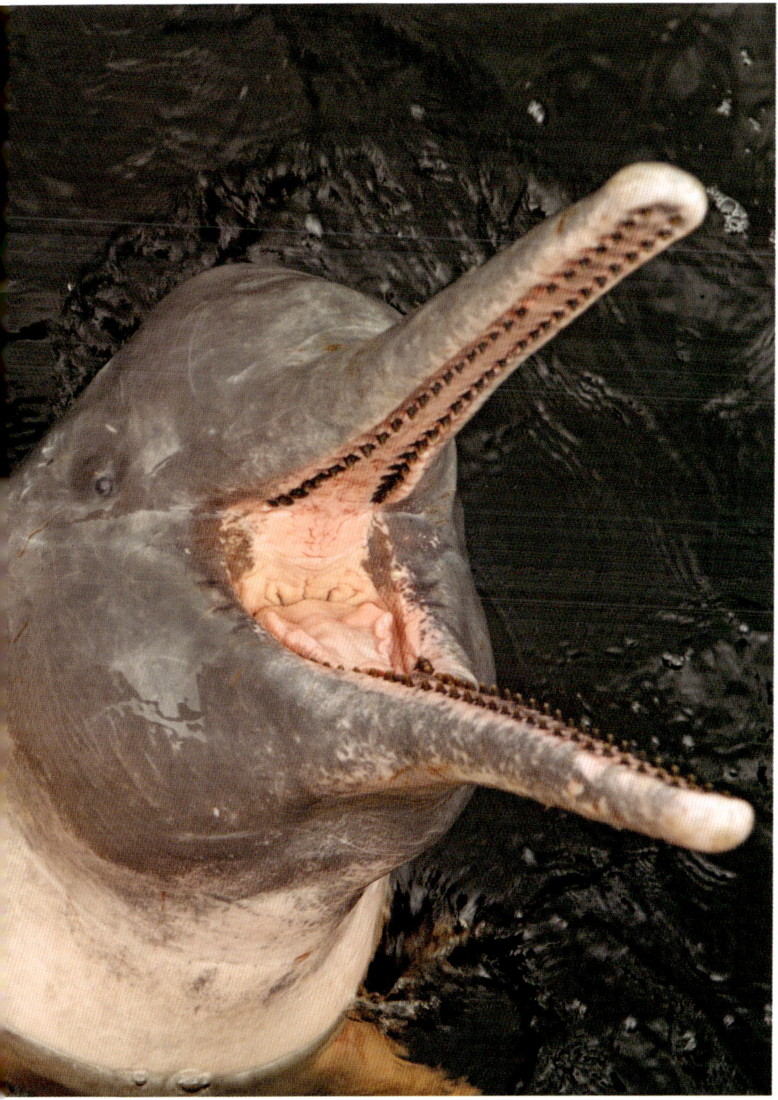

RIGHT TOP AND BOTTOM:

Black-footed ferret

As a predator of the prairie dogs of North America's great plains, this burrowing hunter has suffered as the prairies were cleared to make way for cattle and crops. Today a captive breeding programme is steadily increasing their numbers, and there are thought to be 1,000 in the wild.

OPPOSITE:

Andean cat

This wild mountain cat is only a little larger than a domestic pet. About 1,300 of them survive in the rocky grasslands of the Andes.

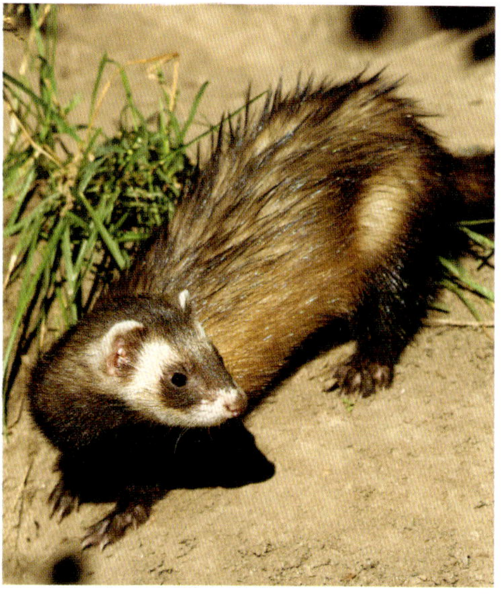

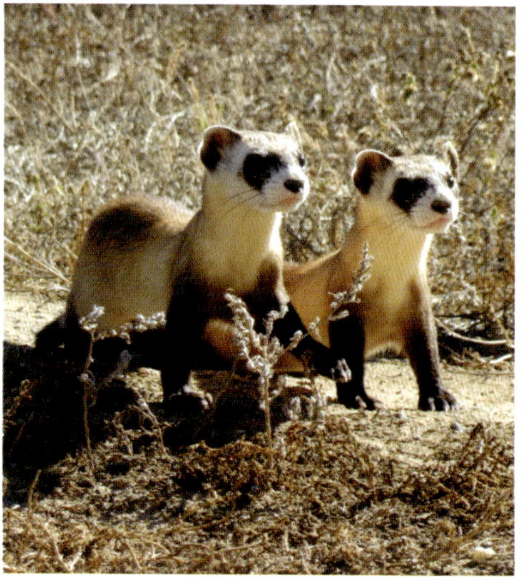

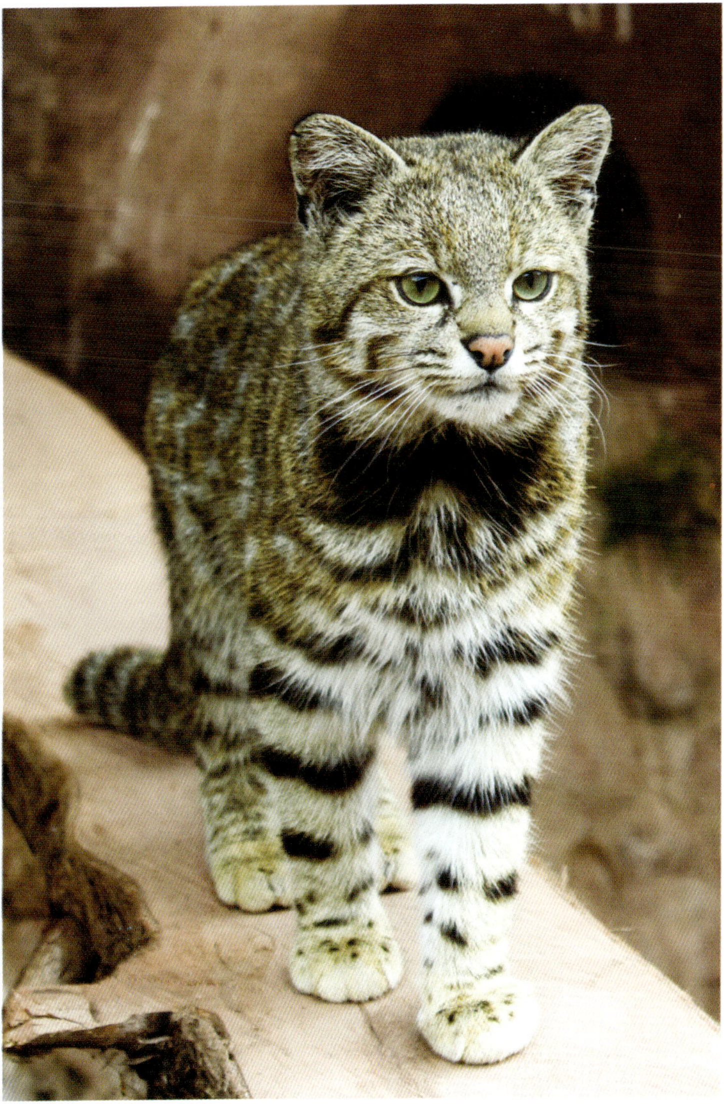

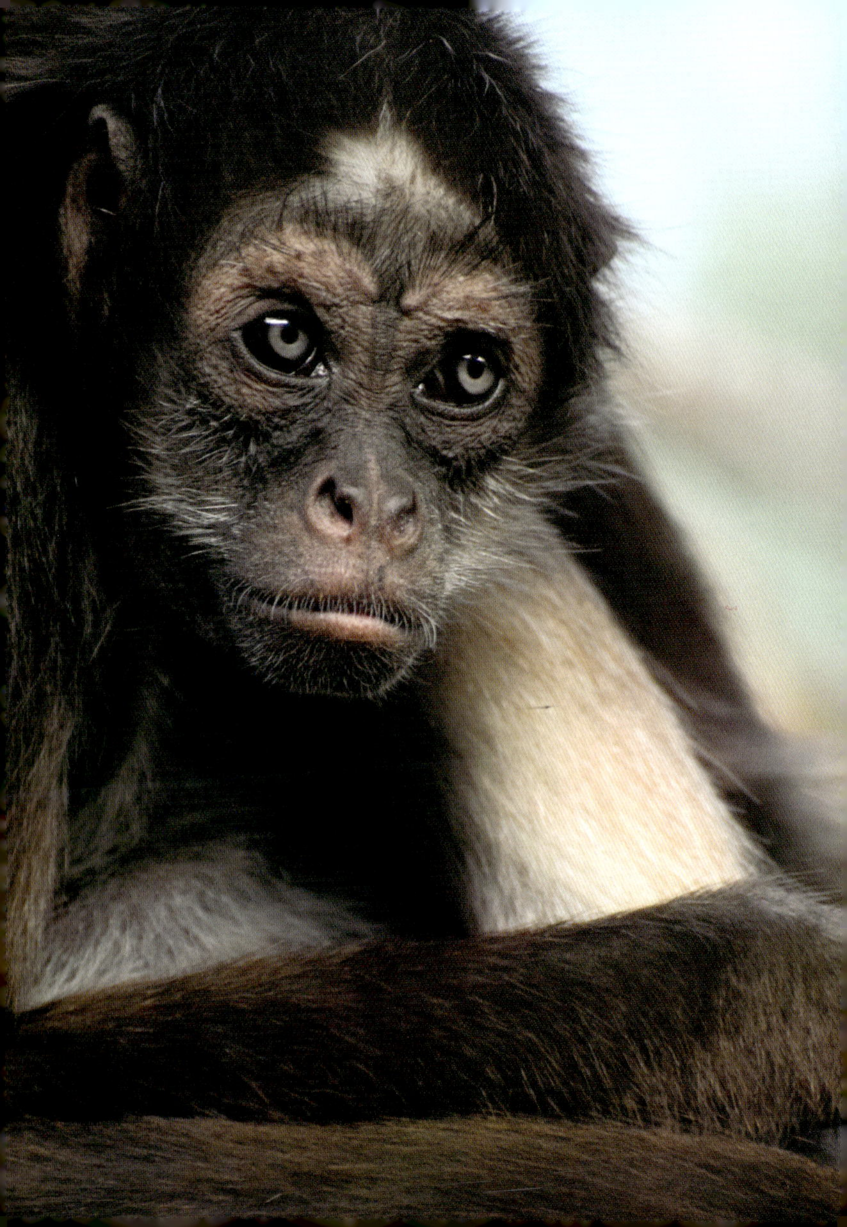

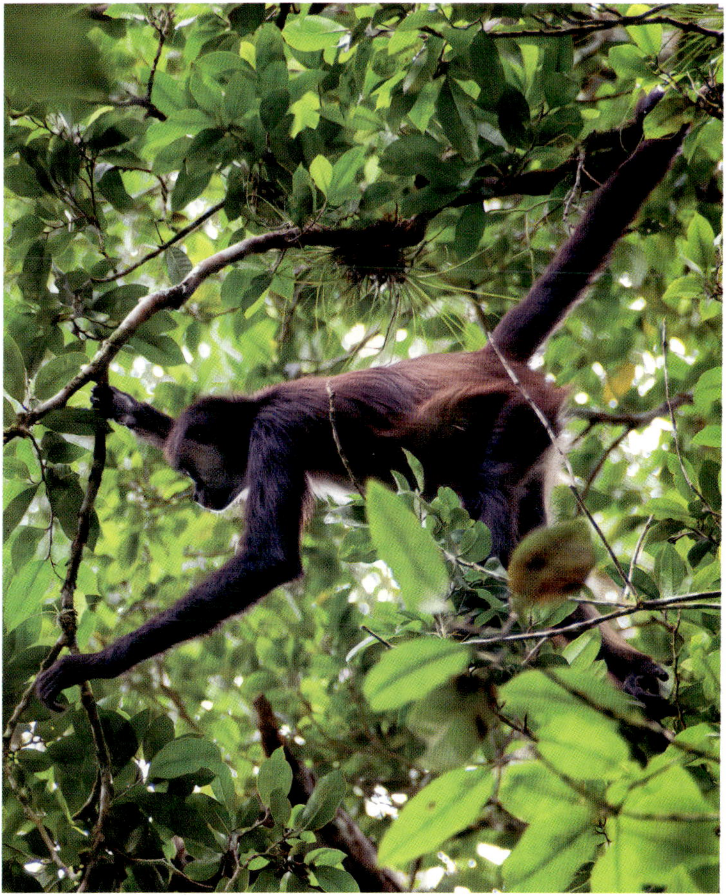

ALL PHOTOGRAPHS:

Brown spider monkey

These monkeys need pristine forest with tall trees and so are under extreme pressure. Probably the rarest and most endangered monkey species in the Americas, it is thought just 30 live wild in Colombia while the rest of the world population is protected in captivity.

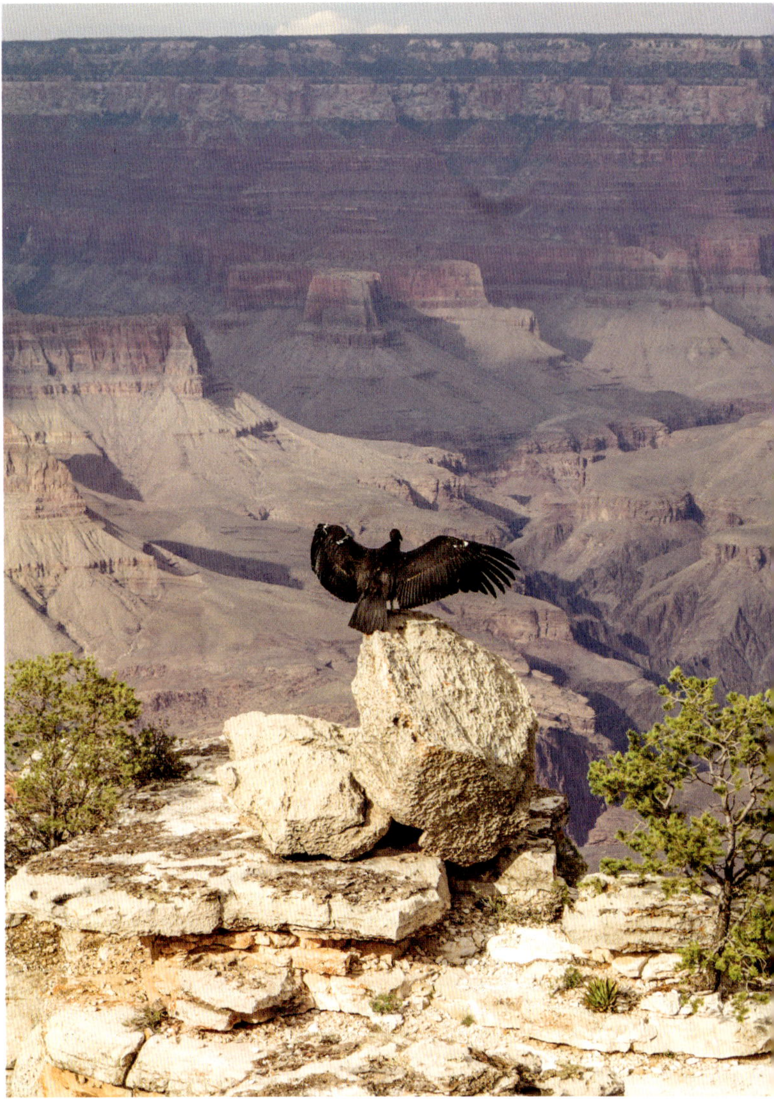

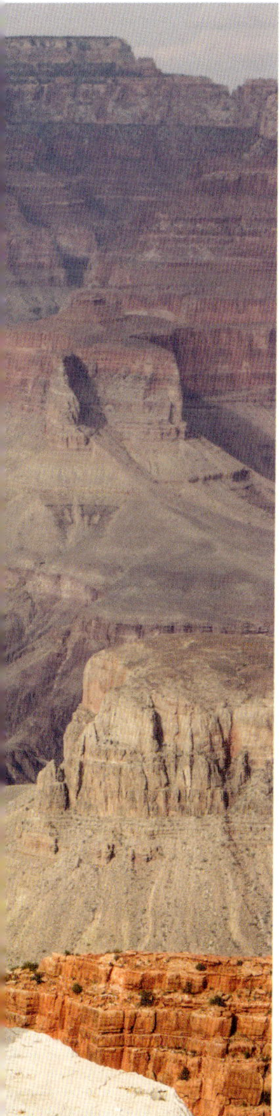

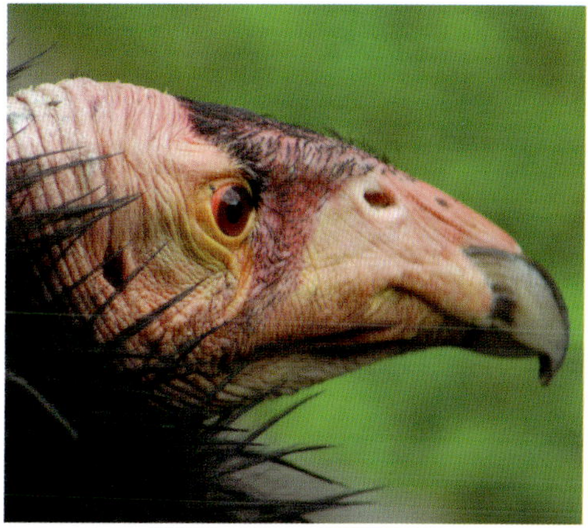

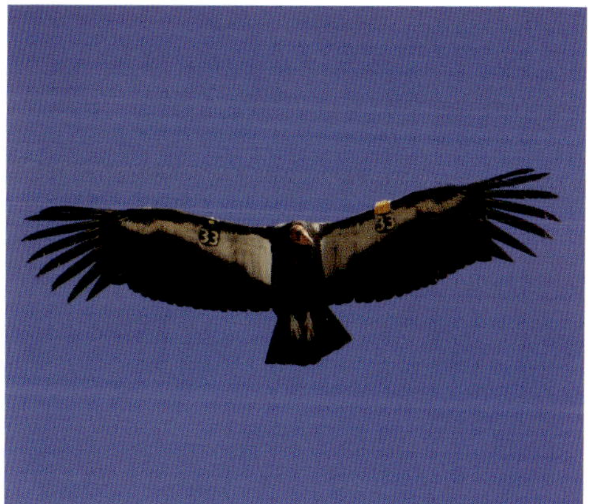

ALL PHOTOGRAPHS:

Californian condor

One of the early conservation success stories, this scavenging bird was on the brink of extinction due to habitat loss and lead poisoning, where the birds ate the remains of animals shot by sports hunters. Today numbers are on the rise again thanks to a captive breeding programme.

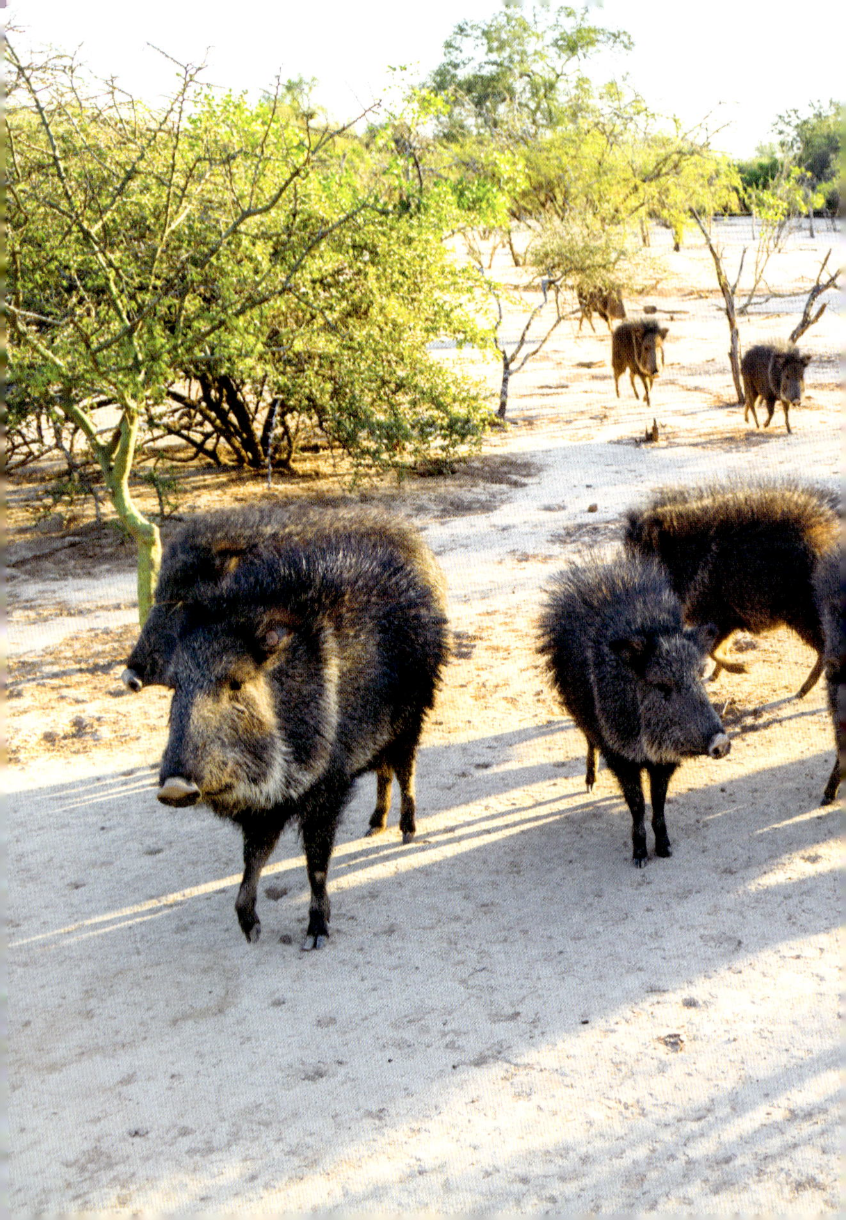

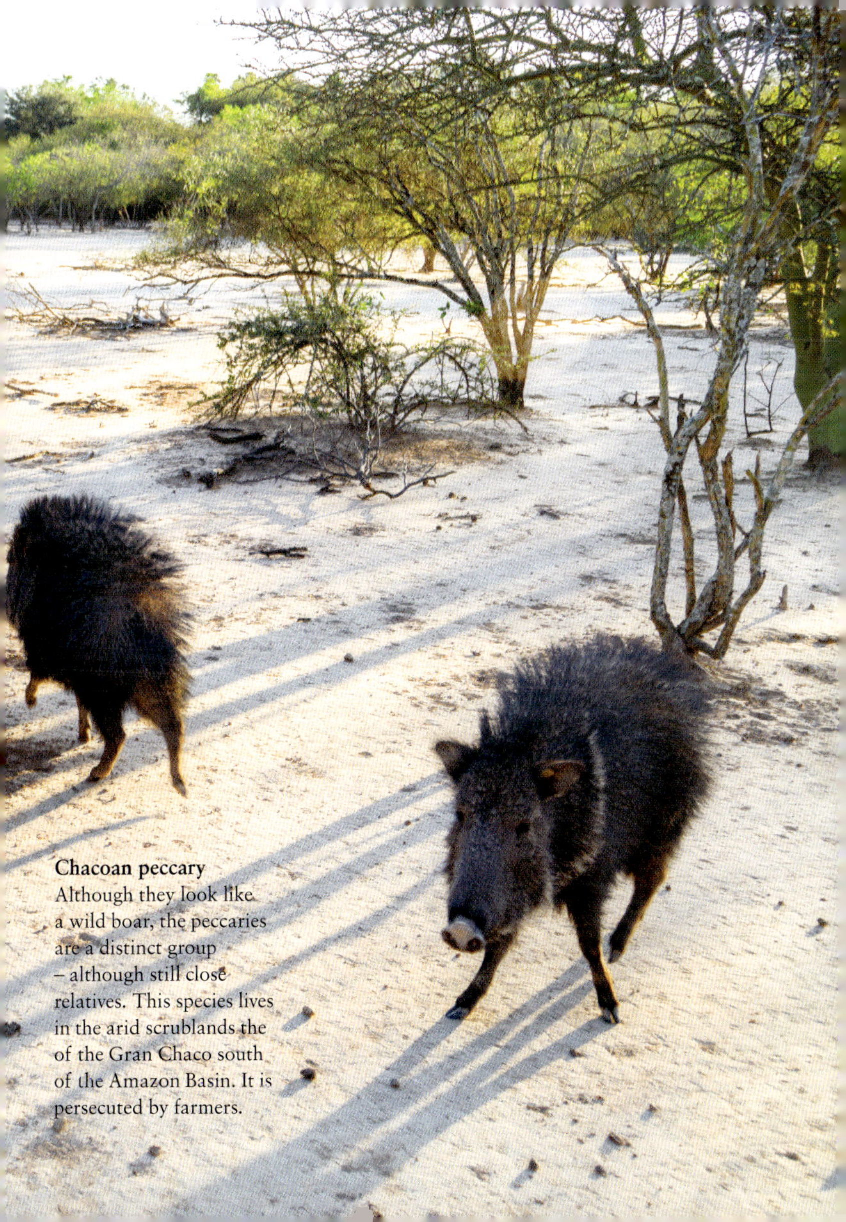

Chacoan peccary
Although they look like a wild boar, the peccaries are a distinct group – although still close relatives. This species lives in the arid scrublands the of the Gran Chaco south of the Amazon Basin. It is persecuted by farmers.

Chinchilla

These mountain rodents from the chilly scree habitats of the Andes are famed for having some of the thickest – and warmest – fur in the animal kingdom. As a result, their numbers have suffered at the hands of fur trappers, and today both species are highly endangered and their numbers continue to drop due to poaching.

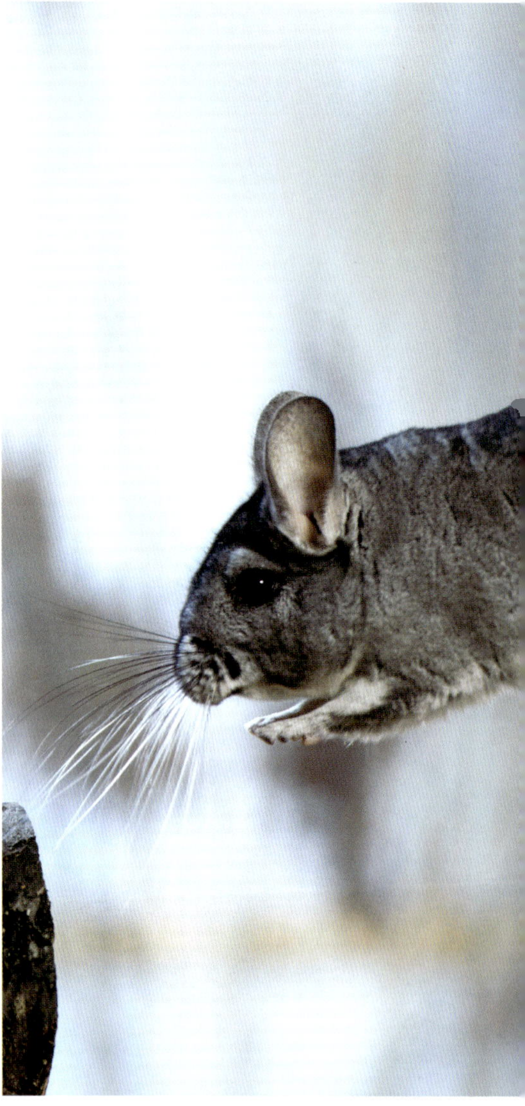

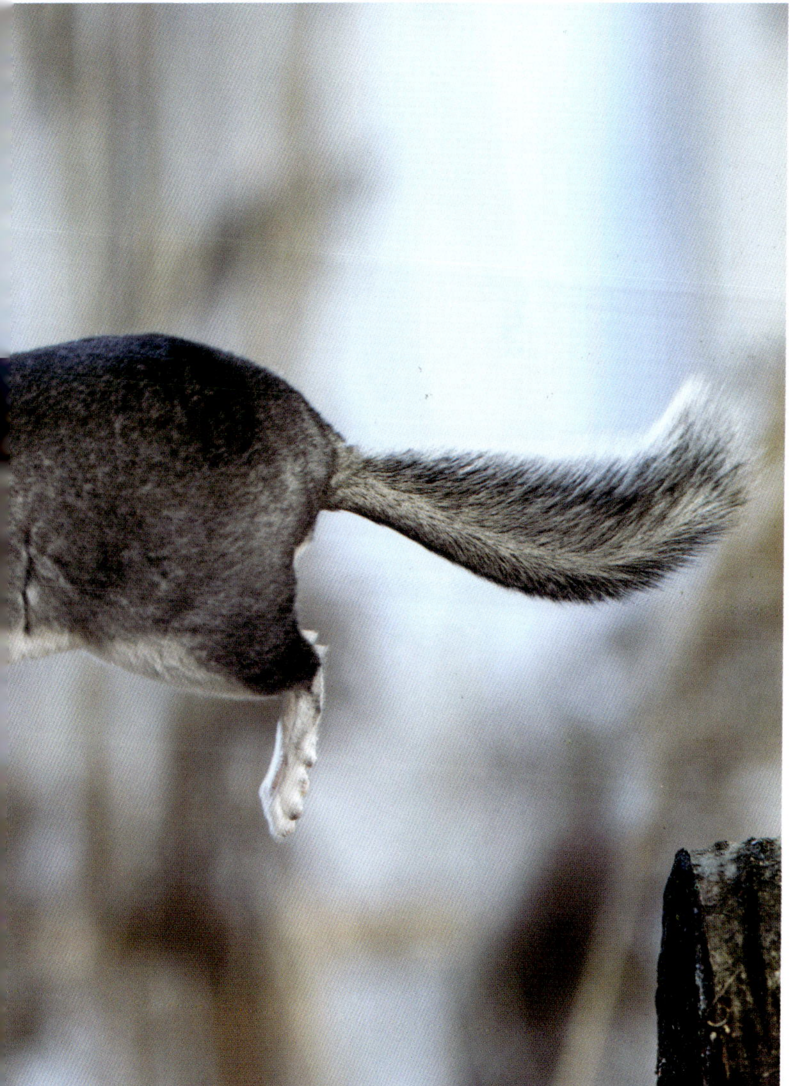

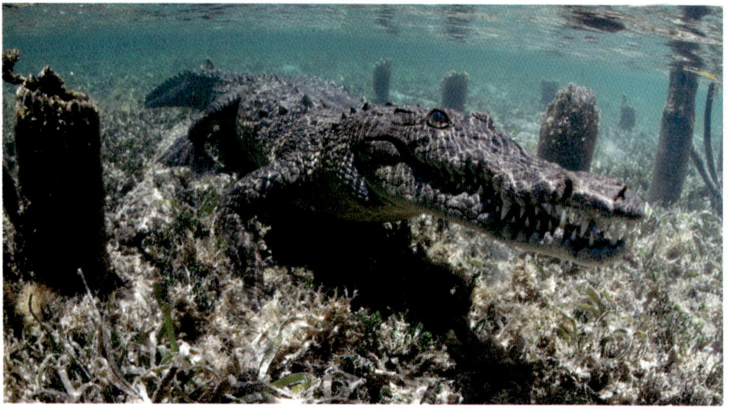

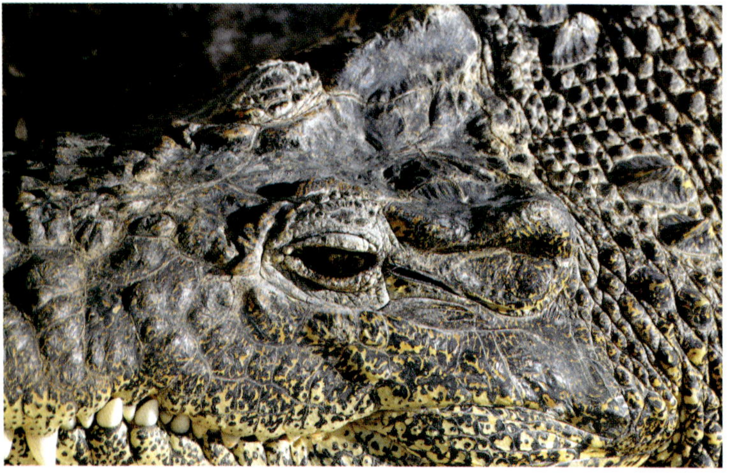

ALL PHOTOGRAPHS:

Cuban crocodile

This small croc is found only in the swamps of Cuba, and is thus highly vulnerable to habitat loss and hunting pressures. Its small wild population is highly protected in nature reserves, where they delight tourists, but numbers remain perilously low.

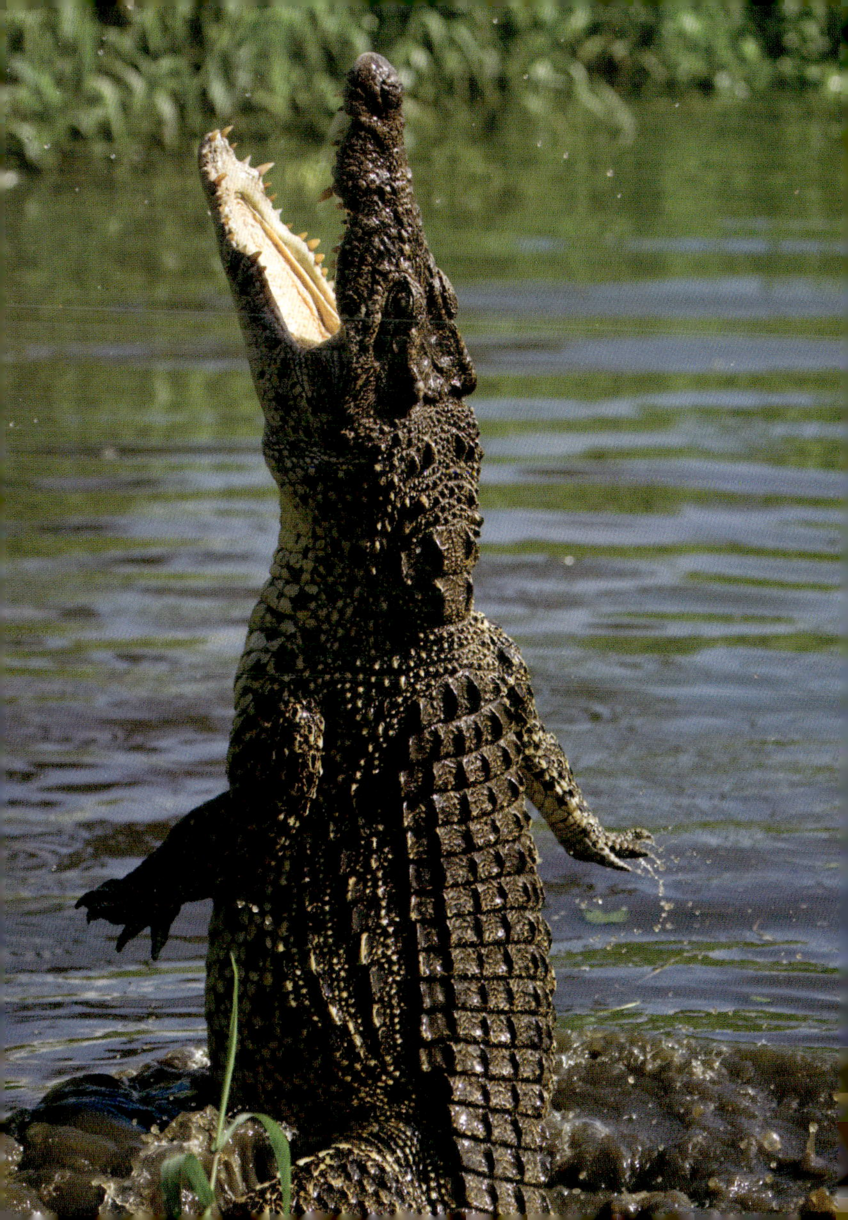

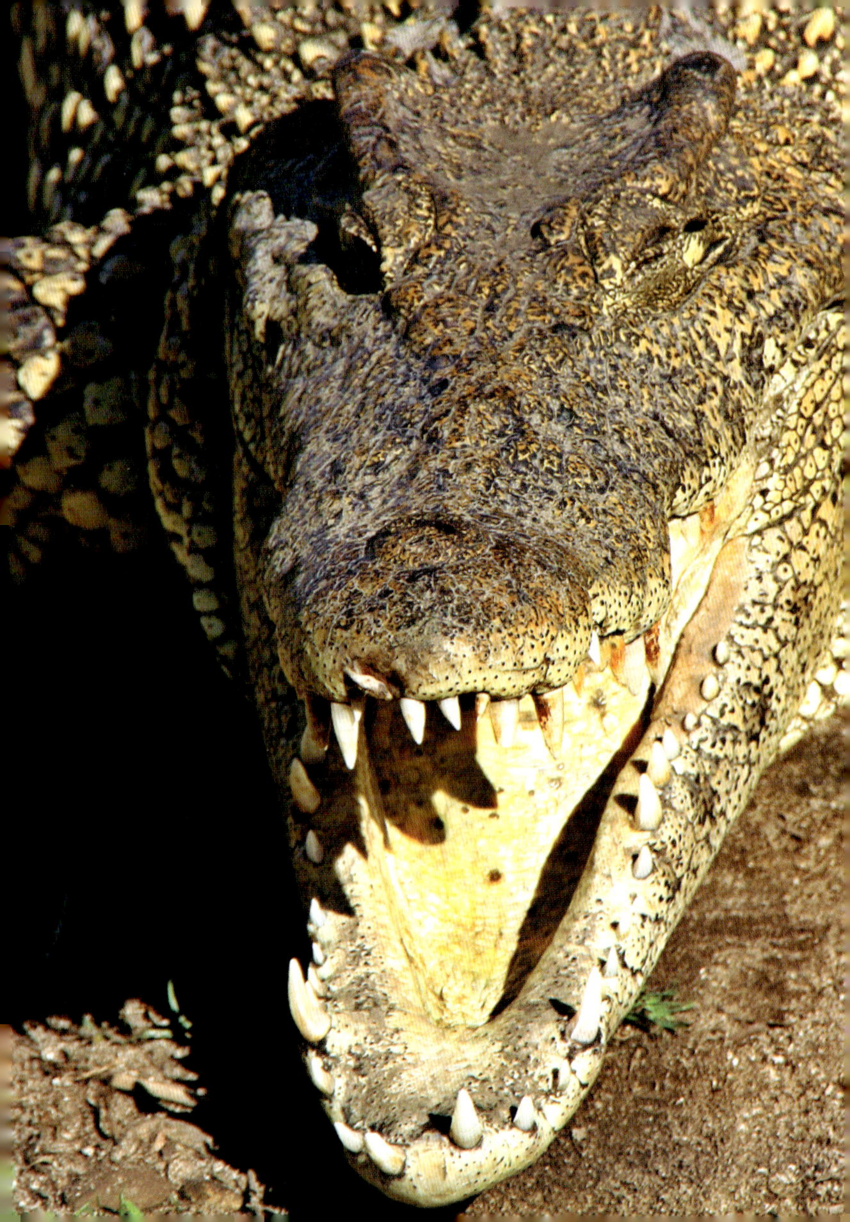

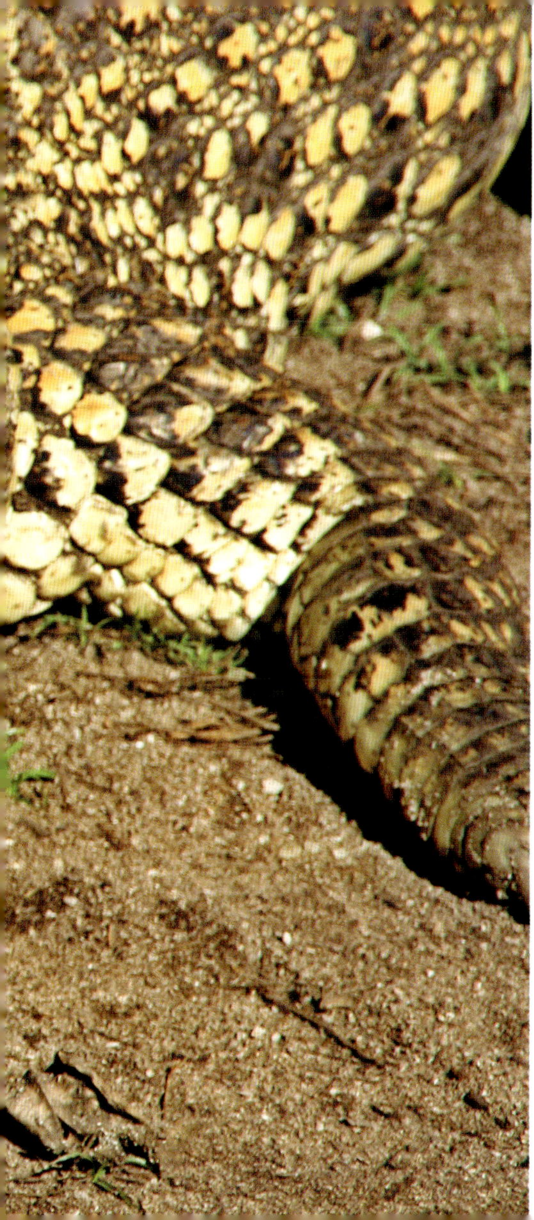

Cuban crocodile

The Cuban crocodile is highly aggressive, more so than the larger American crocodile. The American species lives in both freshwater and marine habitats and so a few visit Cuba's coasts now and then. One of the threats facing the Cuban species is hybridisation with the American visitors.

Darwin's fox

This fox is named after Charles Darwin, who made the first description of the species during his voyage aboard the Beagle in the 1830s. This fox, also known as Darwin's zorro, lives only in patches of forest along the coast of Chile and is threatened by competition with feral dogs.

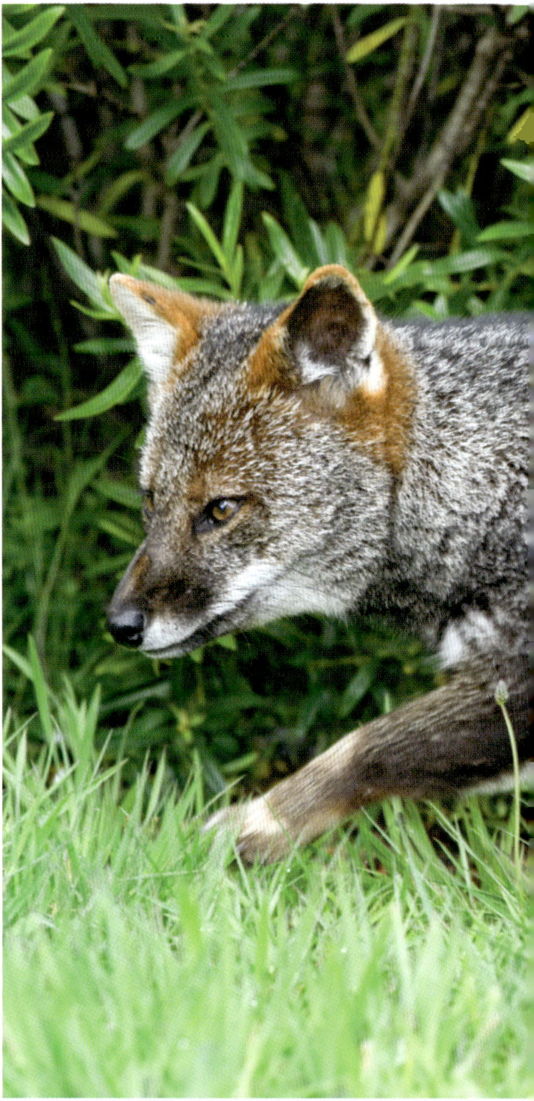

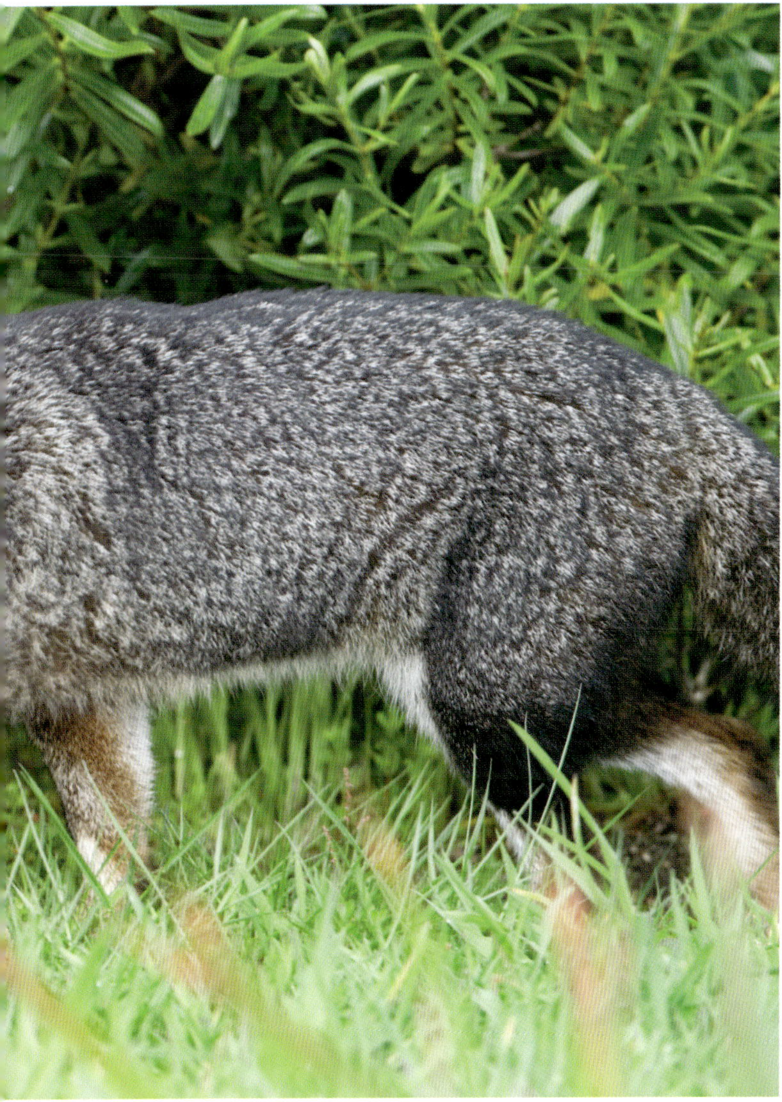

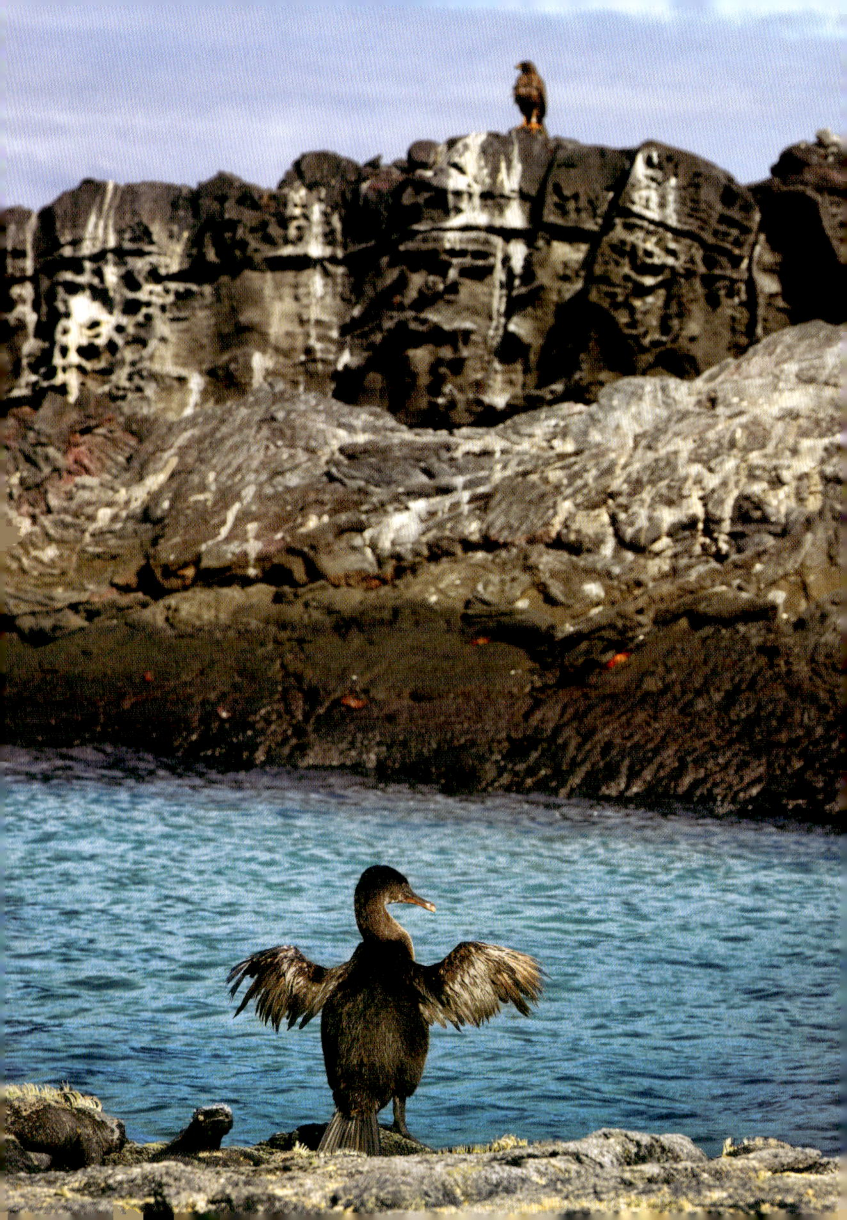

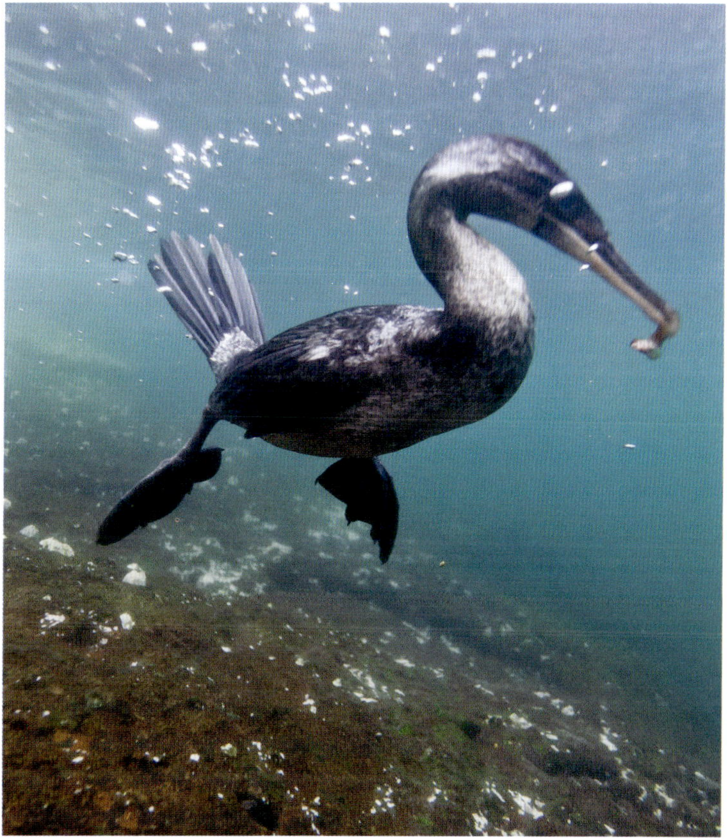

ALL PHOTOGRAPHS:

Flightless cormorant

The cormorants are a widespread group of fish-eating birds that catch and kill prey with their diving skills. The Galápagos Islands host this unique flightless species which dives in from the rocks and has wings too small and weak to fly. The islands are highly protected but the bird is still under threat from the cats, rats and dogs introduced by people.

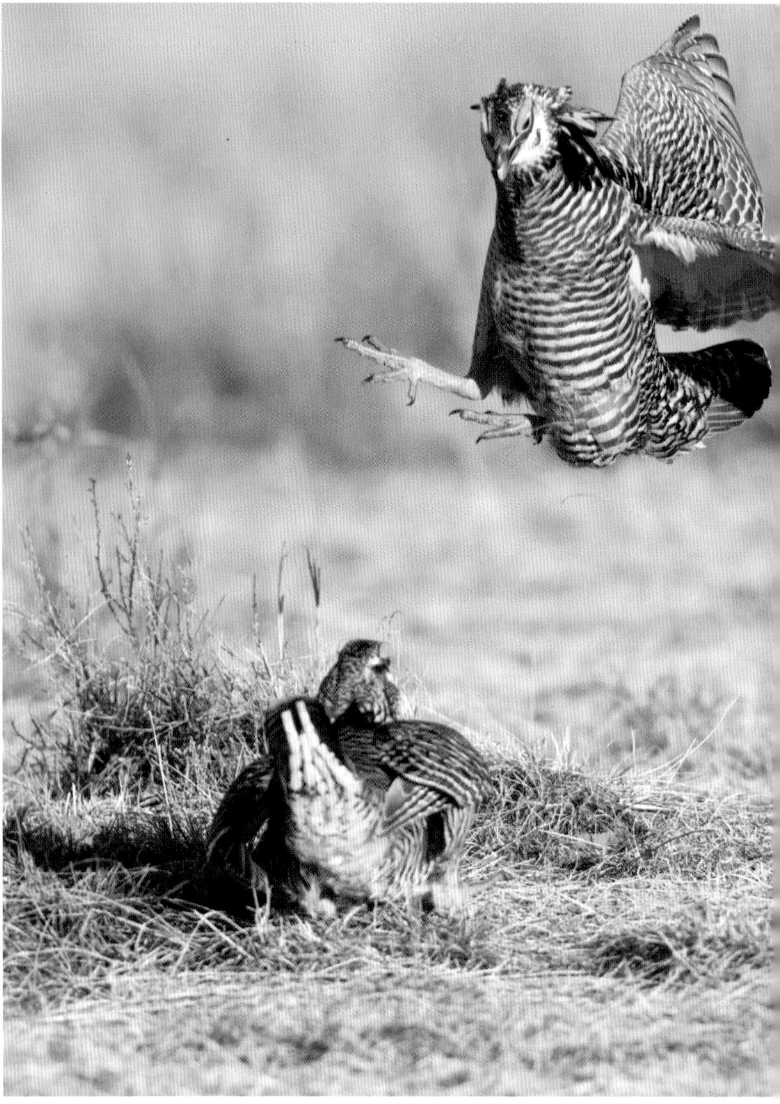

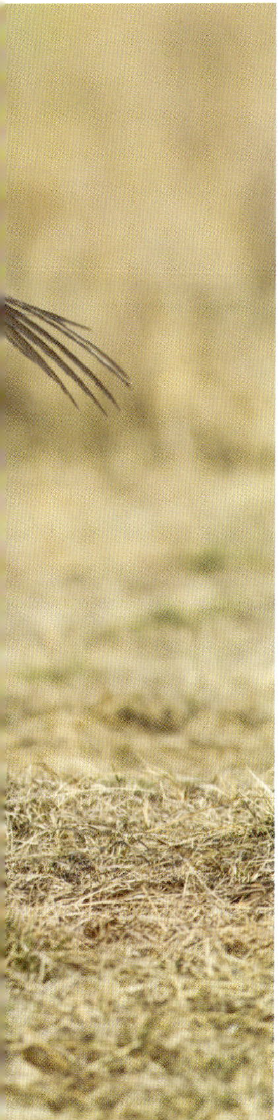

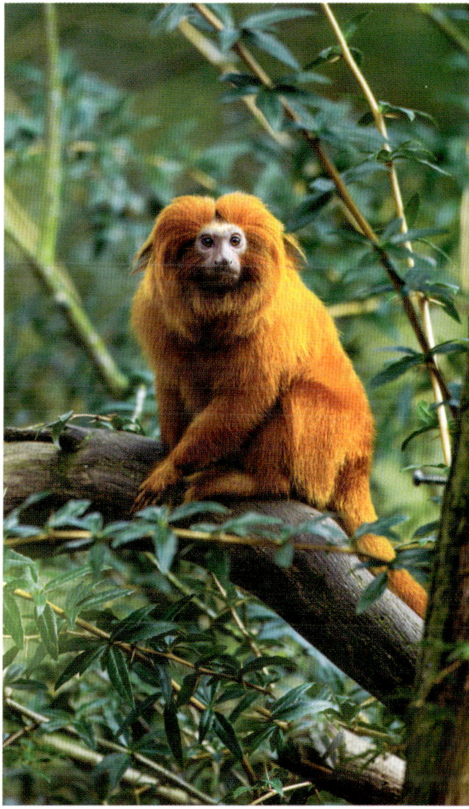

LEFT:

Great prairie chicken
Although still widespread, this North American grouse is under threat from transformation of the vast wild grasslands of the West into farmlands and ranches.

ABOVE:

Golden lion tamarin
This beautiful little monkey from the Atlantic forests of Brazil in the mountains above Rio de Janeiro has been saved by years of captive breeding and reintroduction to protected forest reserves.

Golden lion tamarin
Early reintroductions of the tamarins were a near failure as the captive-bred tamarins were unable to navigate safely through living forests. Many fell to their deaths as the branches swayed beneath their feet. Today zoos keep their tamarins in living trees not in cages with fixed perches.

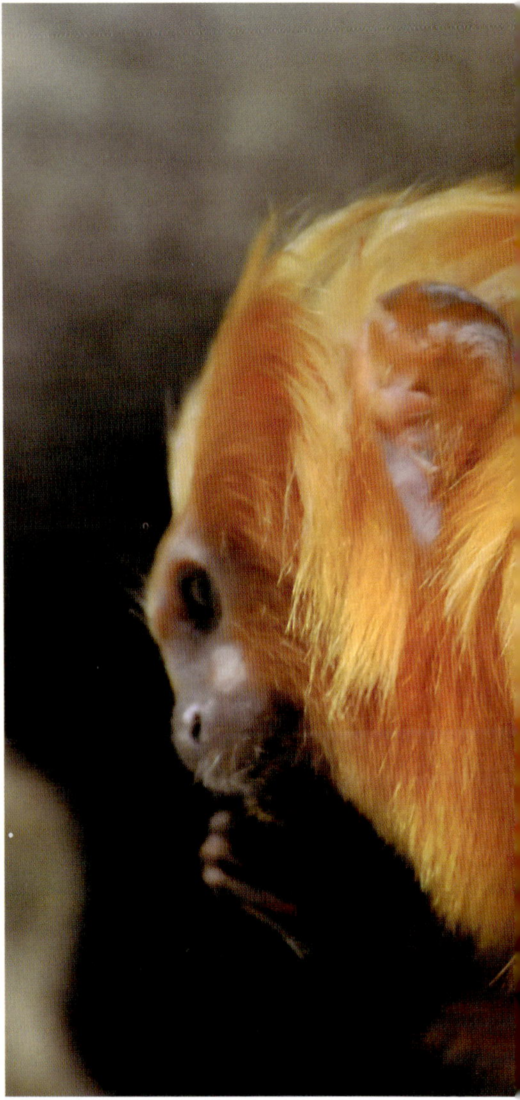

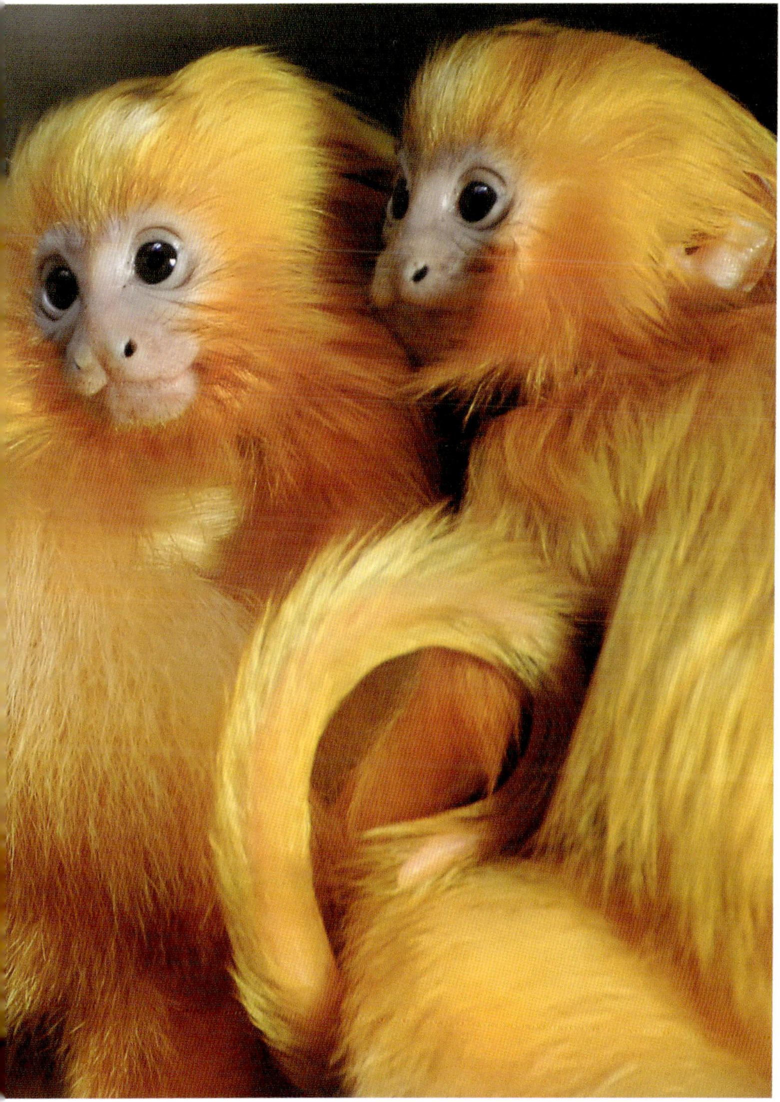

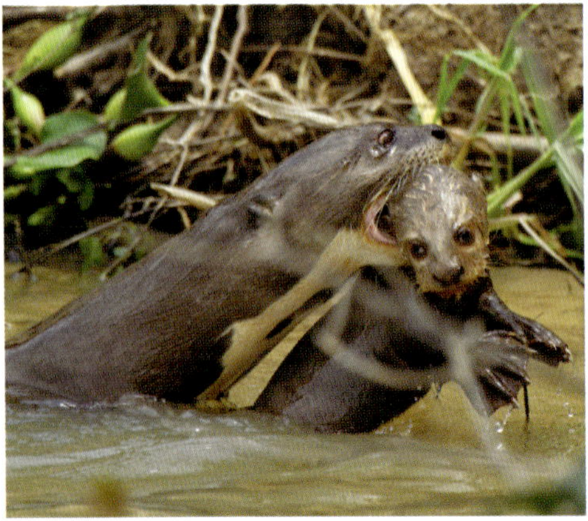

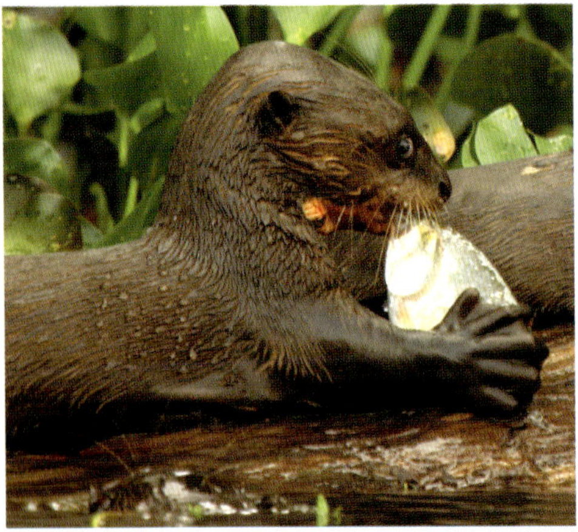

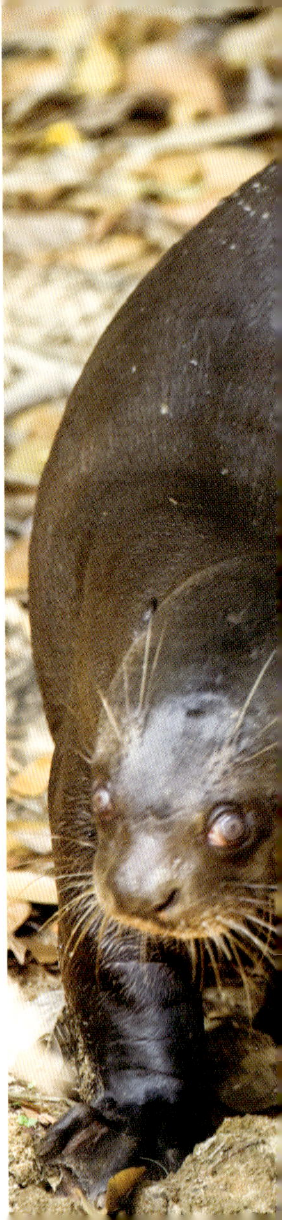

ALL PHOTOGRAPHS:

Giant otter

At 1.7m (5.6ft) long this hunter from the Amazon and Guinean rainforests is by far the world's largest otter. The lure of large pelts led to these creatures being hunted close to extinction. In 1971 there were just 12 left! Today numbers have recovered but the otters are still under threat from habitat loss and pollution.

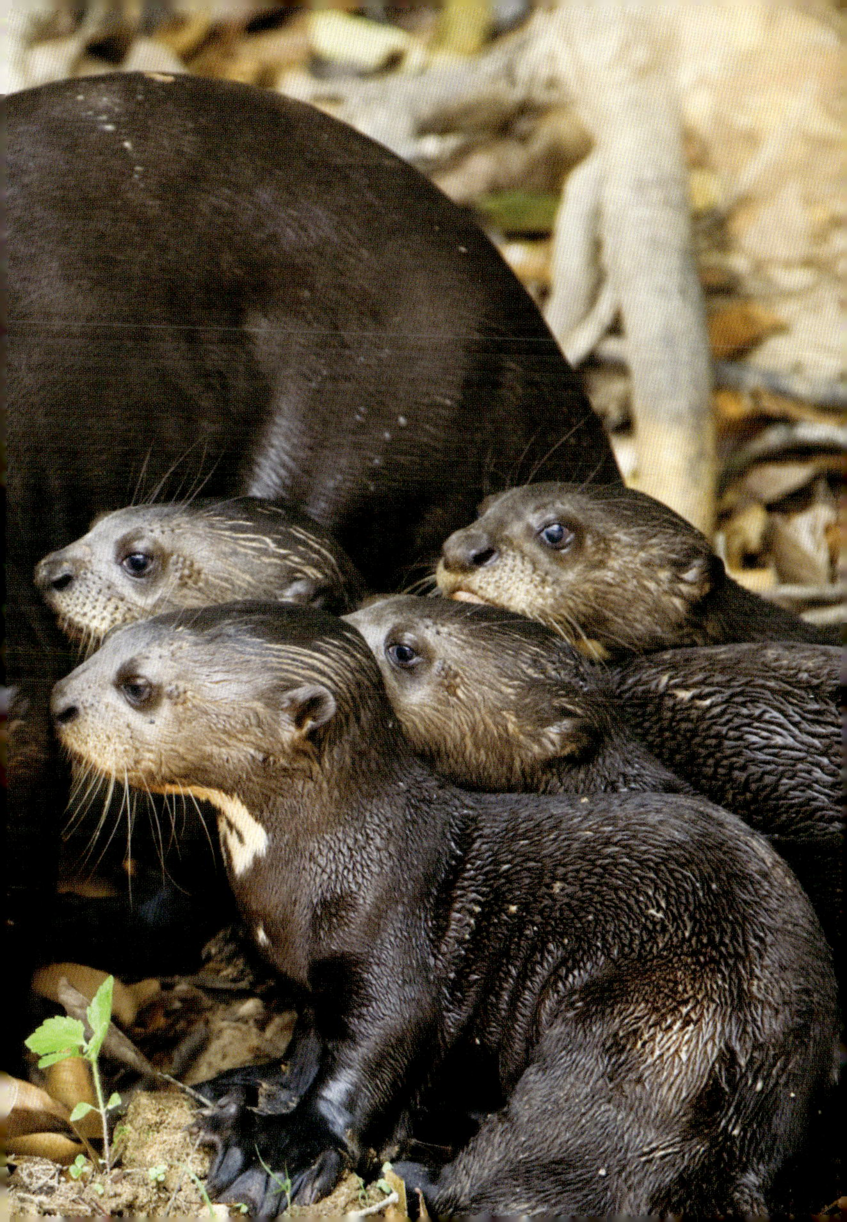

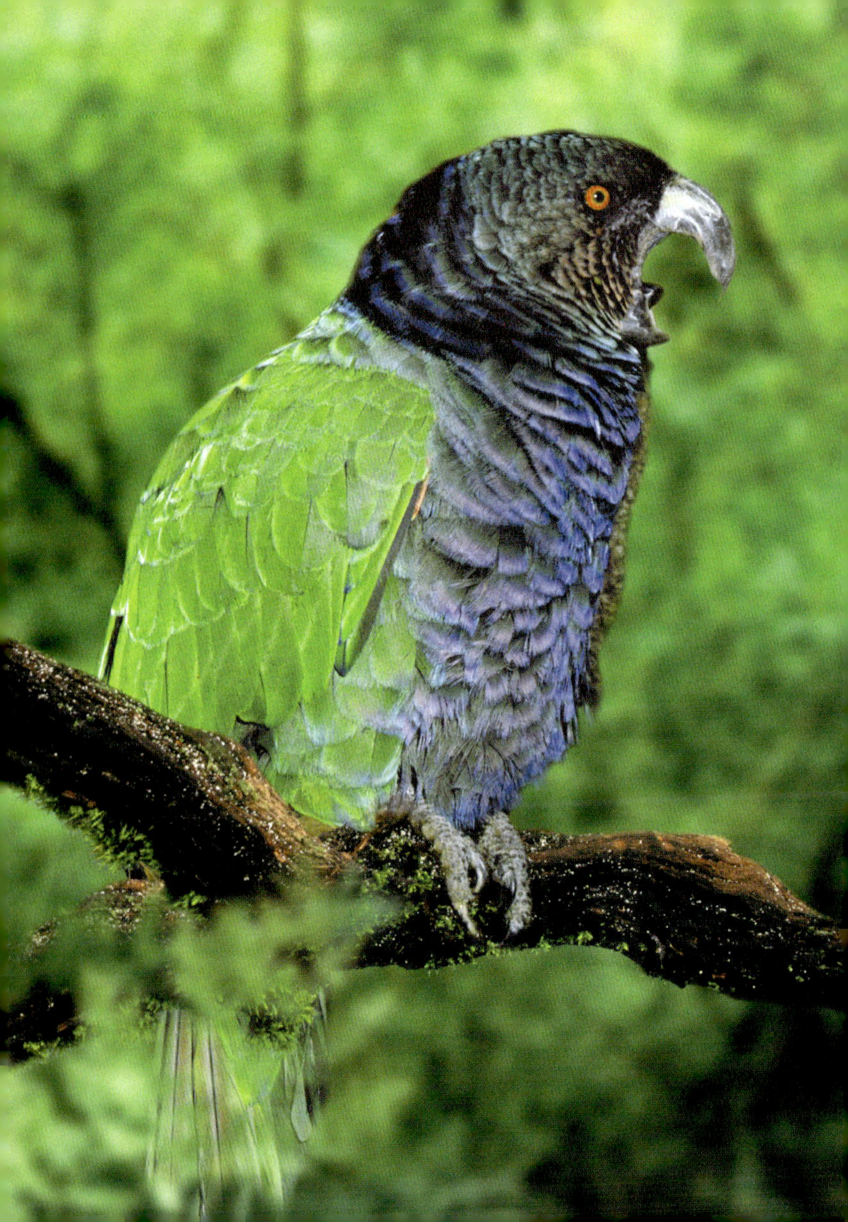

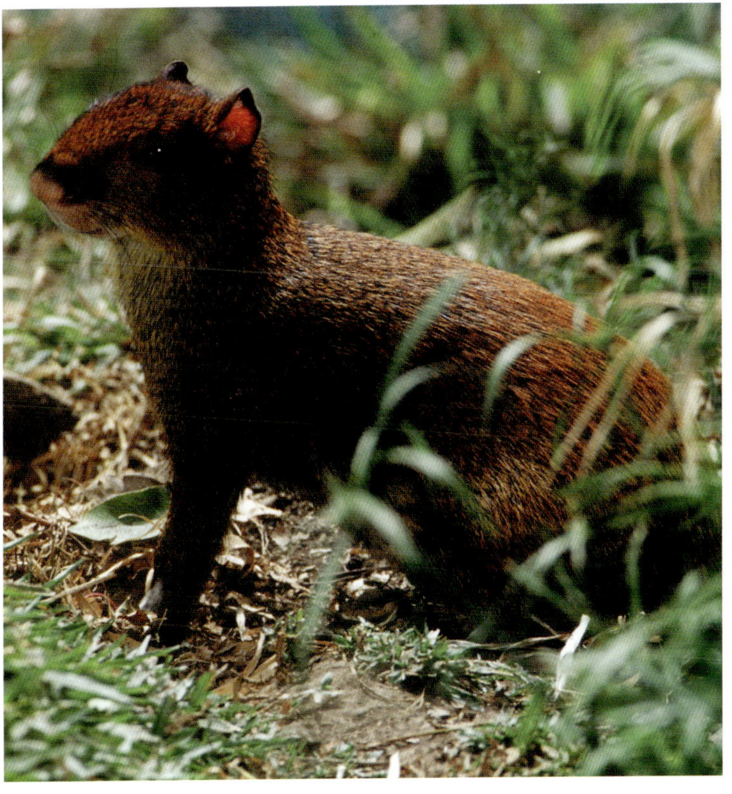

OPPOSITE:

Imperial amazon

This large parrot is at great risk because it lives only x the Caribbean island of Dominica. With only 50 left in the island's forests, the bird is critically endangered.

ABOVE:

Mexican agouti

This large rainforest rodent is found only in a small patch of southern Mexico and areas of Cuba. Other agoutis, which all live in burrows and eat fruits and nuts, are found across tropical America.

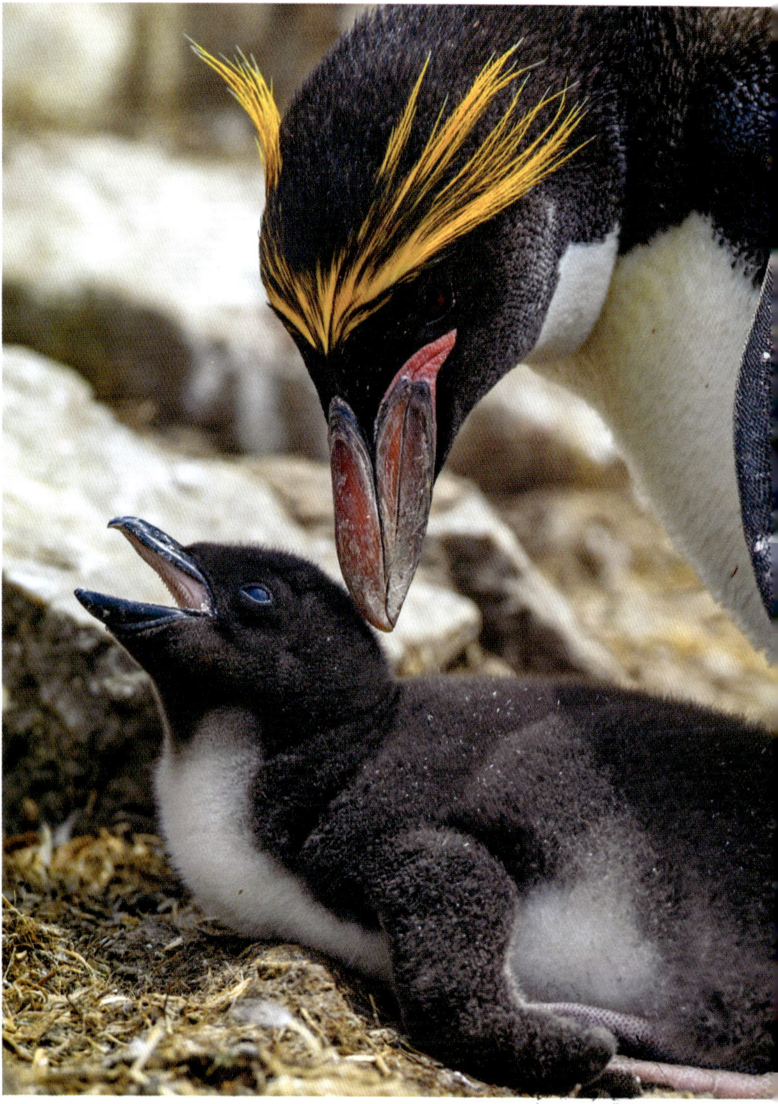

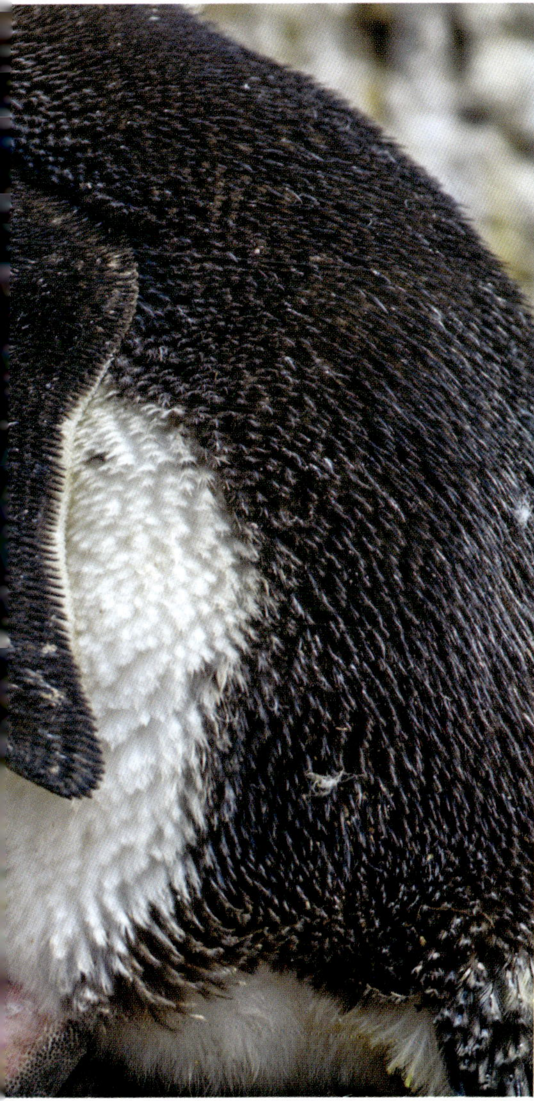

Macaroni penguin

Taking its name from a 19th-century slang word for an exuberantly dressed young man, this penguin lives around Cape Horn and islands across the Southern Ocean. Its global population is in the millions but it is in decline. There are frequent and mysterious population crashes which are a cause for concern.

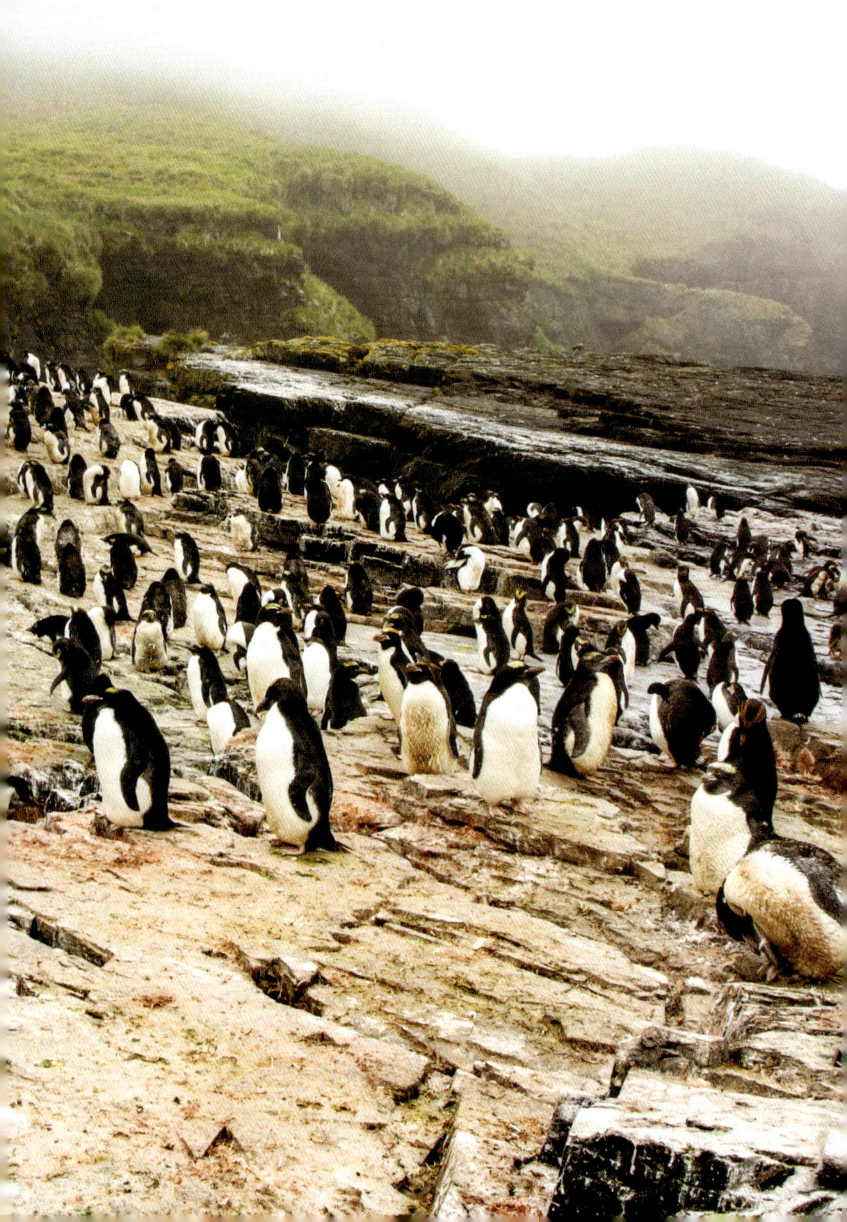

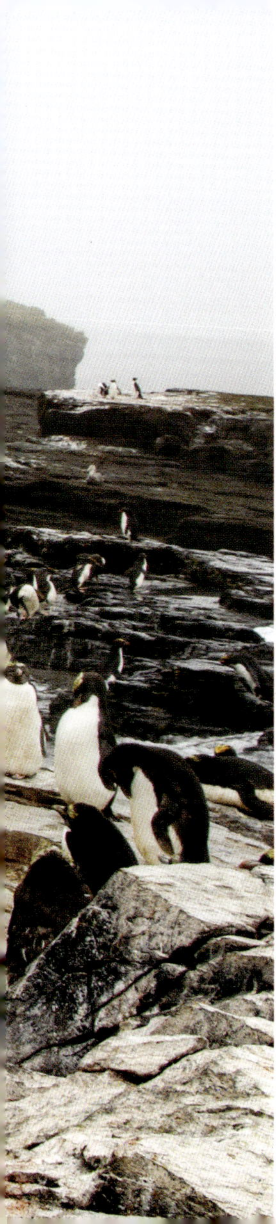

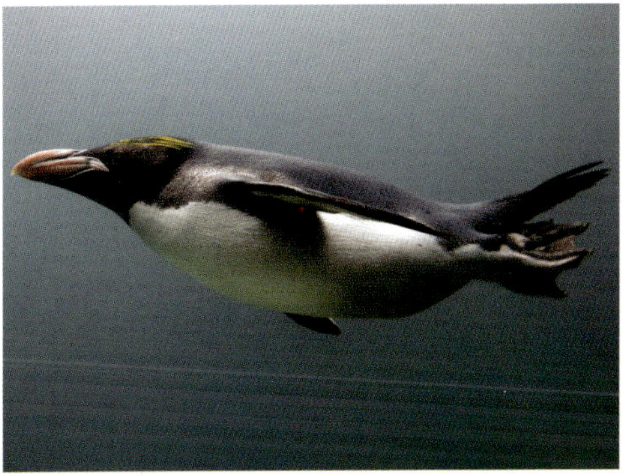

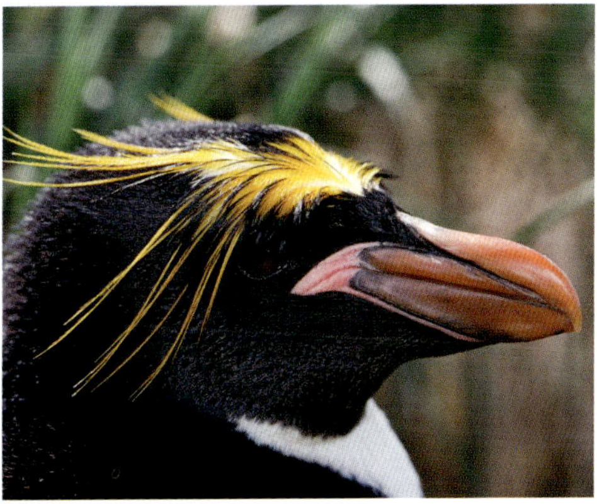

ALL PHOTOGRAPHS:

Macaroni penguin

A leading theory into the apparent decline of this species is that climate change is altering the fish available in the Subantarctic, which makes it harder for mothers to fatten up prior to laying eggs.

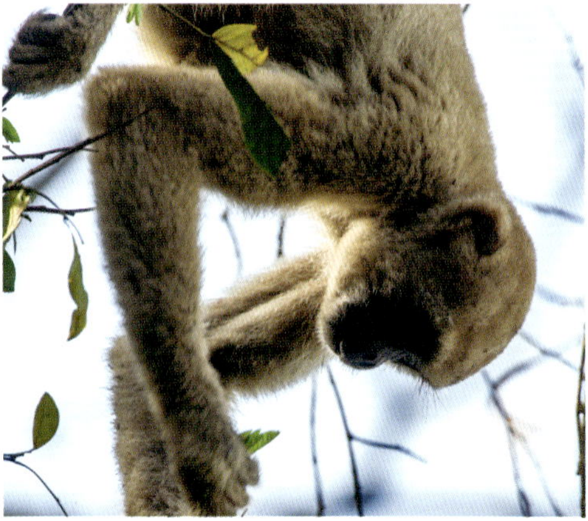

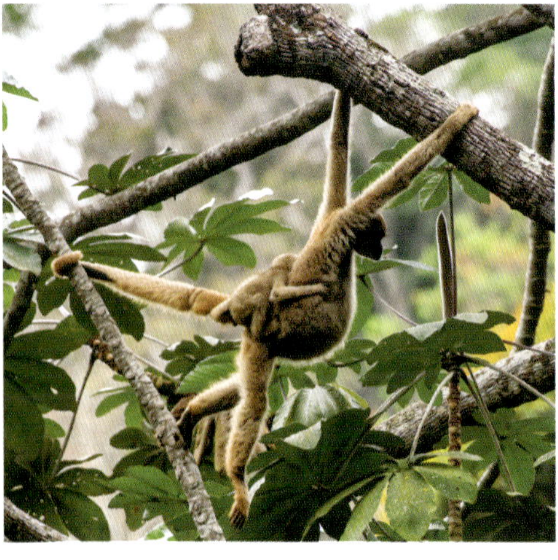

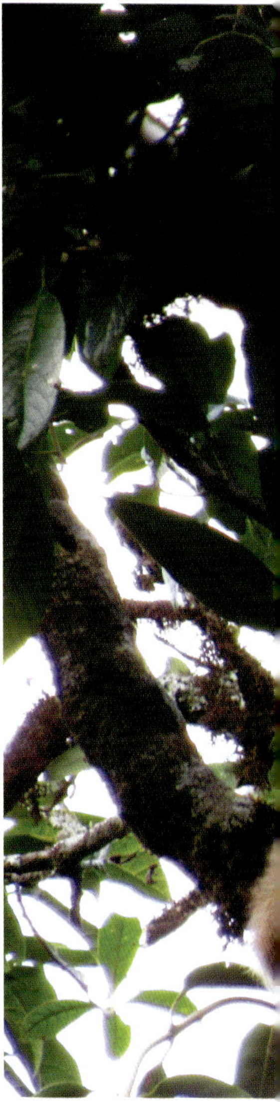

ALL PHOTOGRAPHS:

Muriqui

Also known as woolly spider monkeys, the two muriqui species are the largest monkeys in the Americas – and both are critically endangered. They live in the Atlantic forest of Brazil and are at risk from hunters and habitat fragmentation.

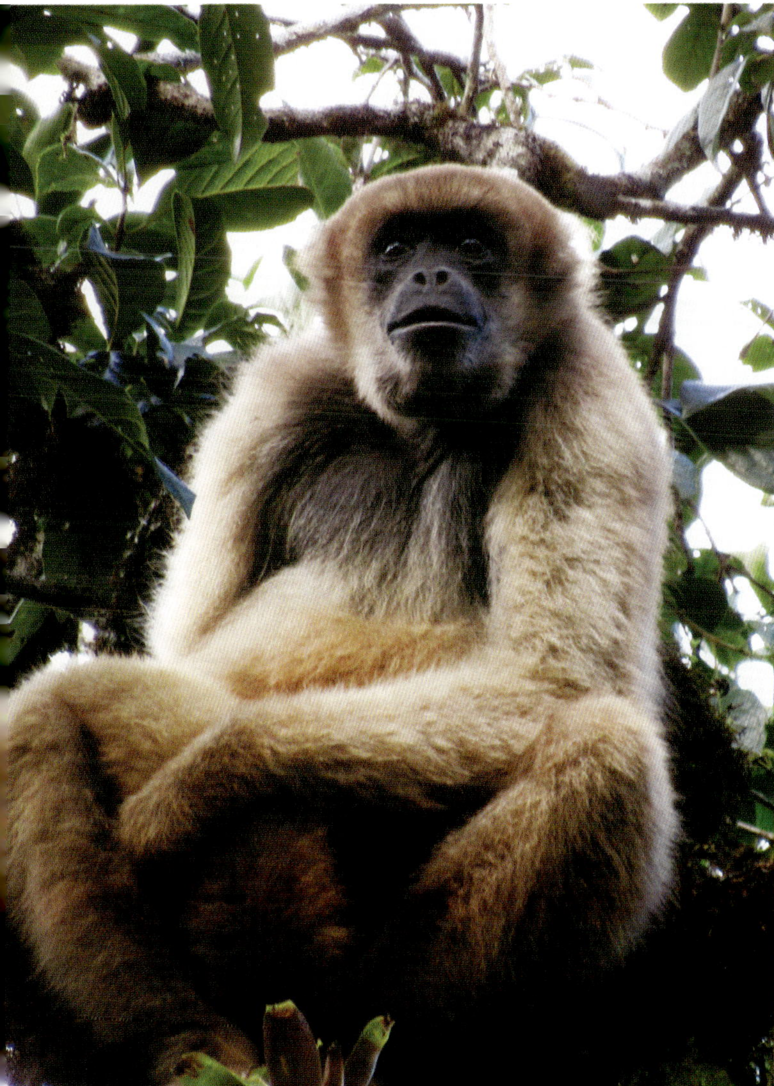

Northern pudu

The world's smallest deer, growing only 35cm (14in) to the shoulder, this little herbivore lives in the high forests of the Andes. These habitats are in decline, and no one is sure how many northern pudus remain there.

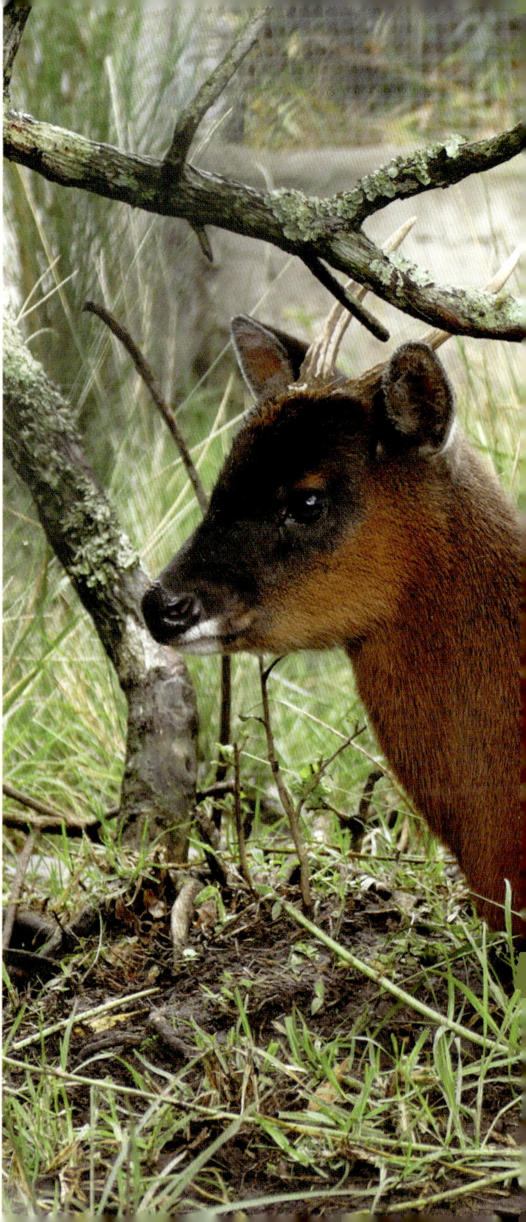

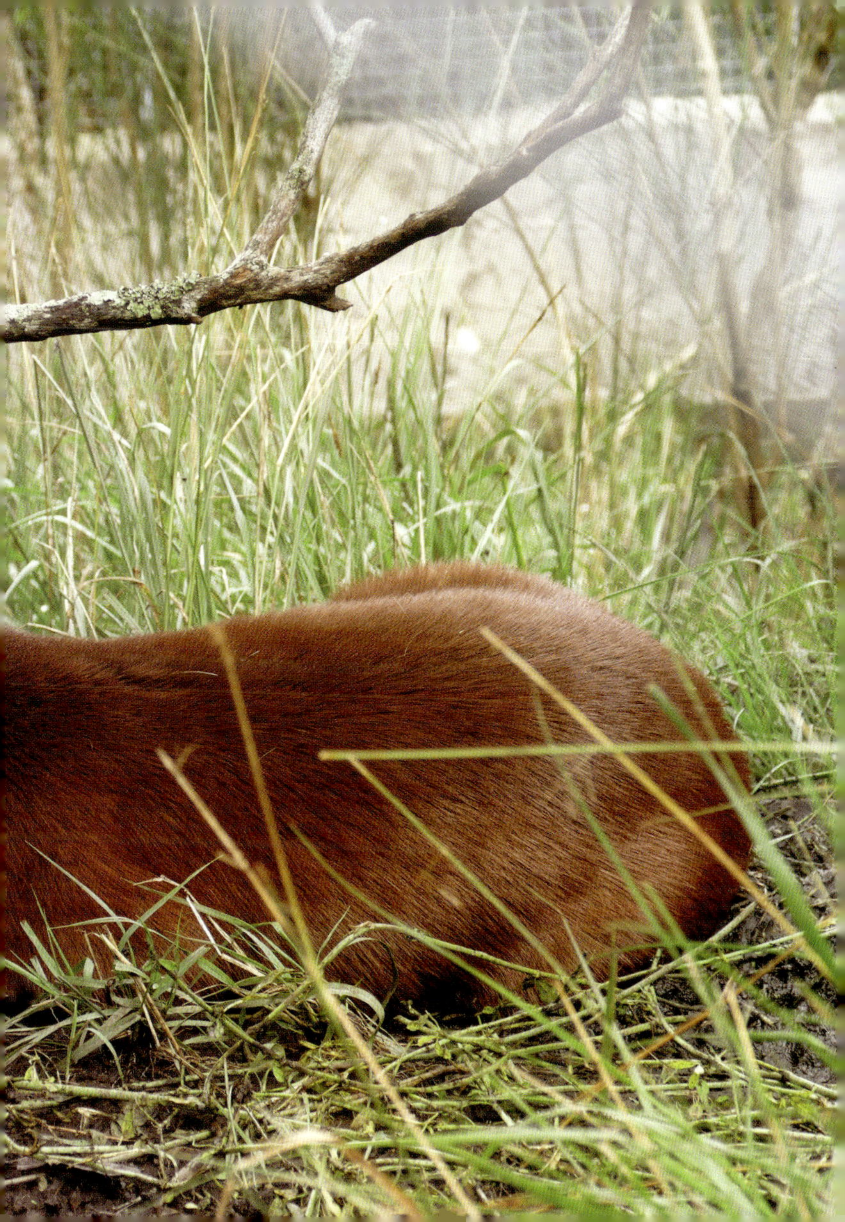

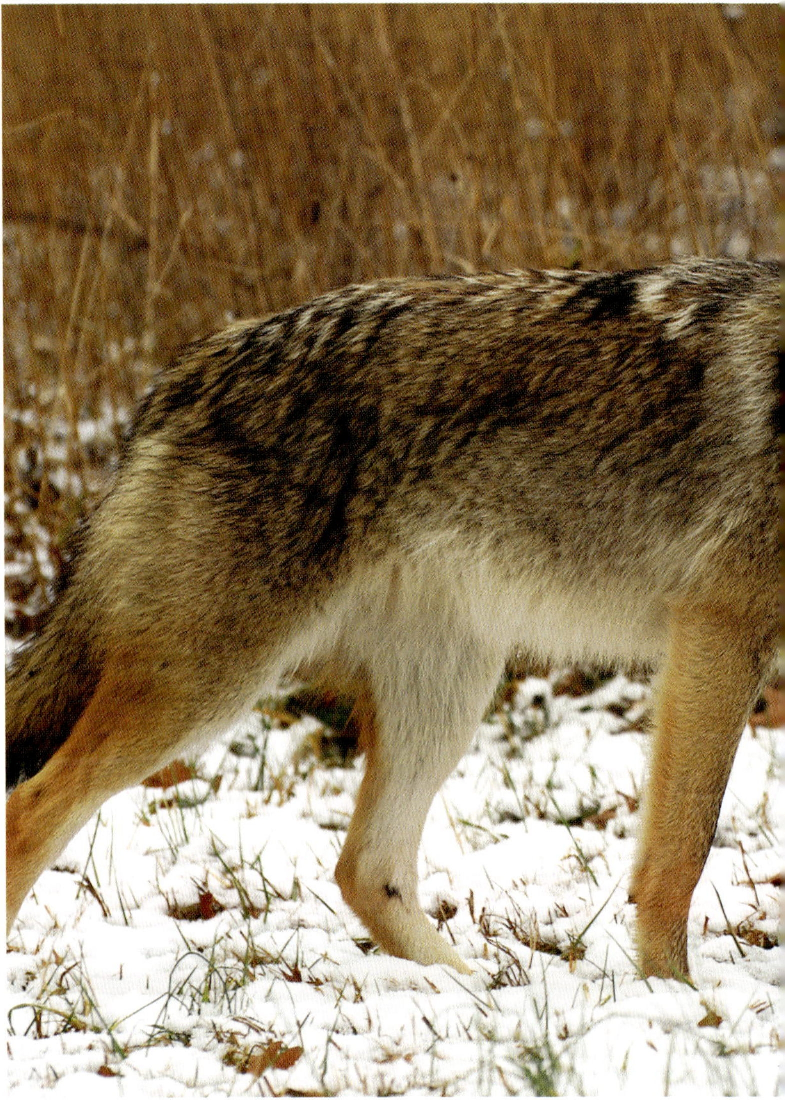

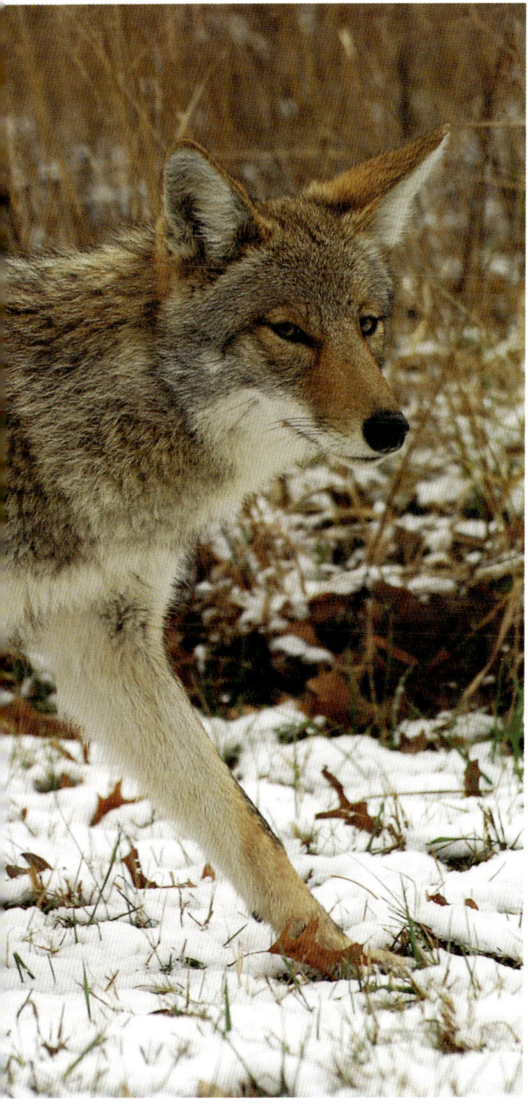

LEFT AND ALL PHOTOGRAPHS OVERLEAF:

Red wolf

The grey wolf is one of the most successful and widespread animal species on Earth. By contrast, its smaller close relative, the red wolf, probably became extinct in the wild in the 1960s. It has since been reintroduced to a protected area in North Carolina, USA.

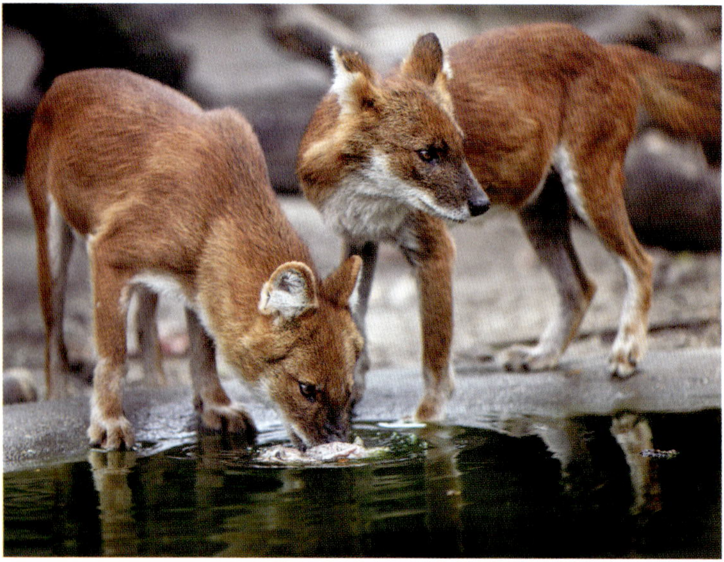

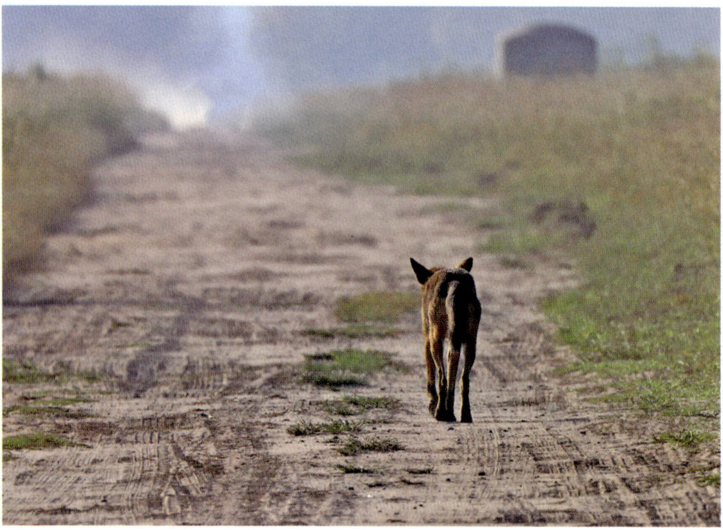

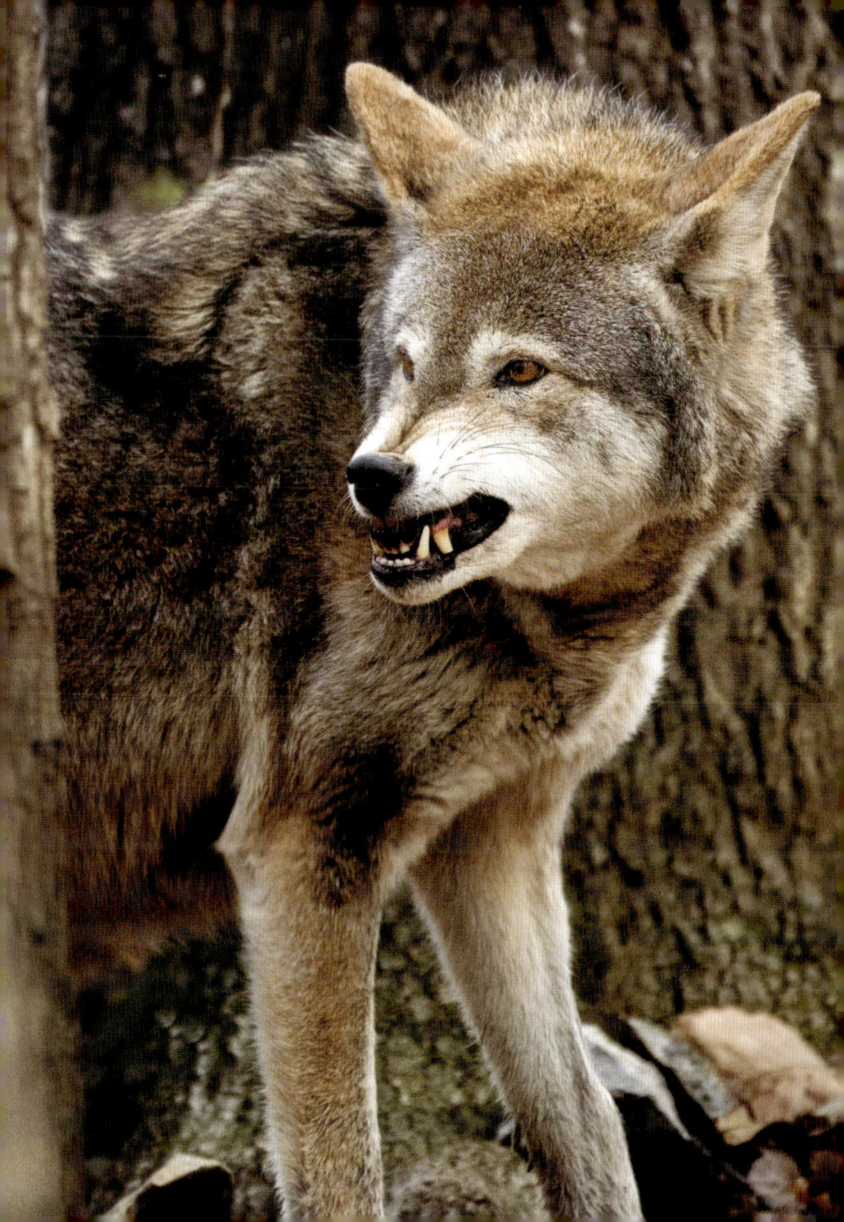

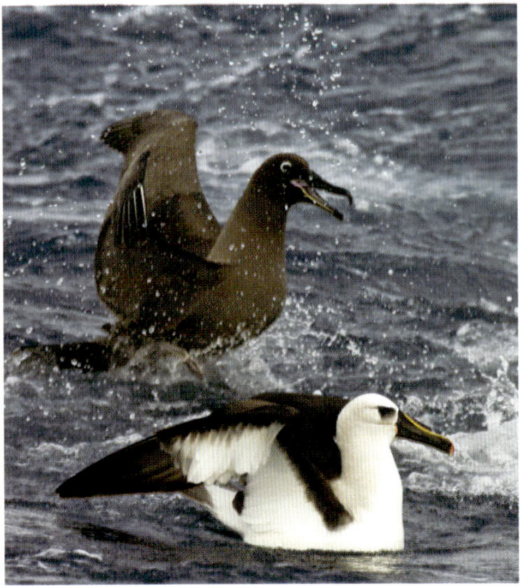

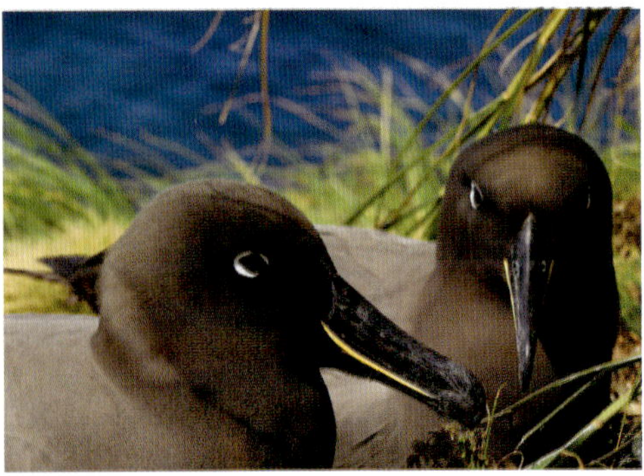

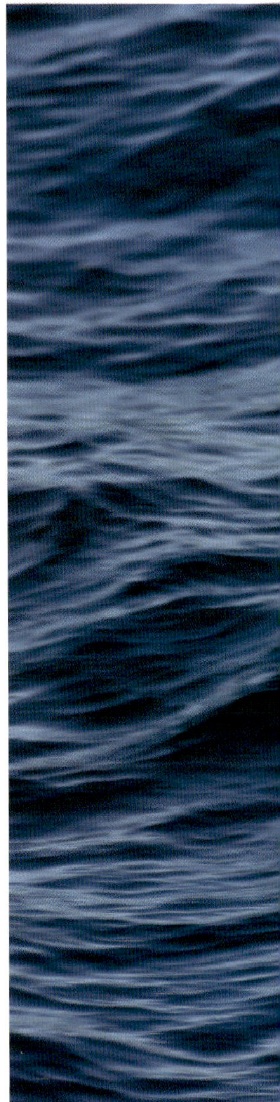

ALL PHOTOGRAPHS:

Sooty albatross

Much darker in colour than other albatrosses (although the young birds have white feathers), this species lives across the islands of the Southern Ocean. It is endangered because its population has reduced by 75 per cent over the last century.

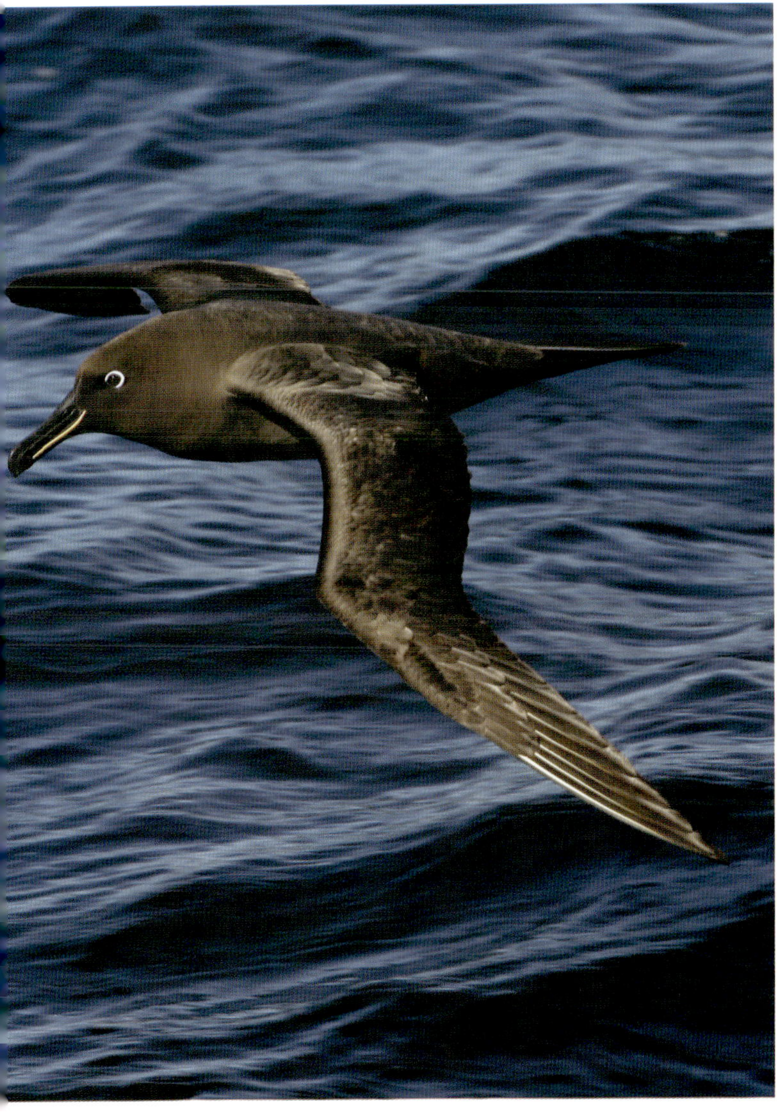

RIGHT:

Vancouver Island marmot

In the 1990s, this ground squirrel was the rarest mammal on Earth with fewer than 30 living in the wild. A breeding programme has since boosted that number to more than 200.

OPPOSITE:

Spectacled bear

The largest mammal in South America, this bear lives a solitary life in the high forests of the Andes. As well as being under pressure from habitat loss, the bear has long been persecuted by local people.

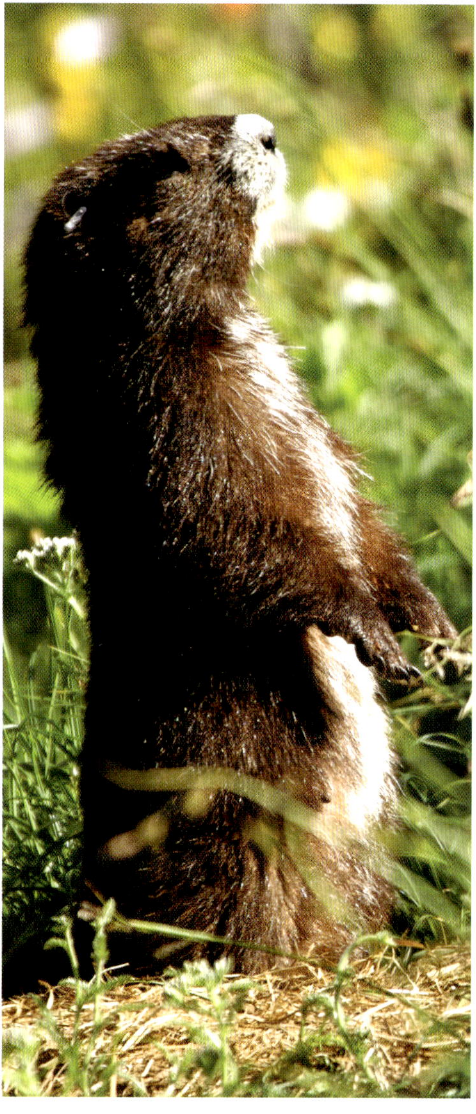

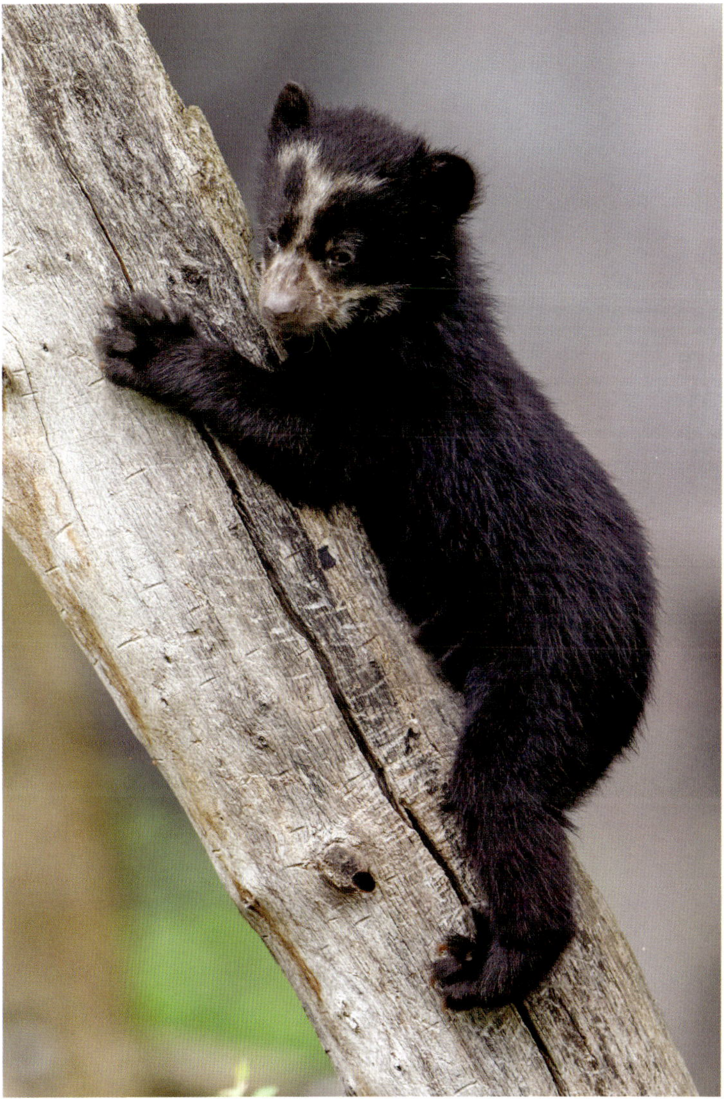

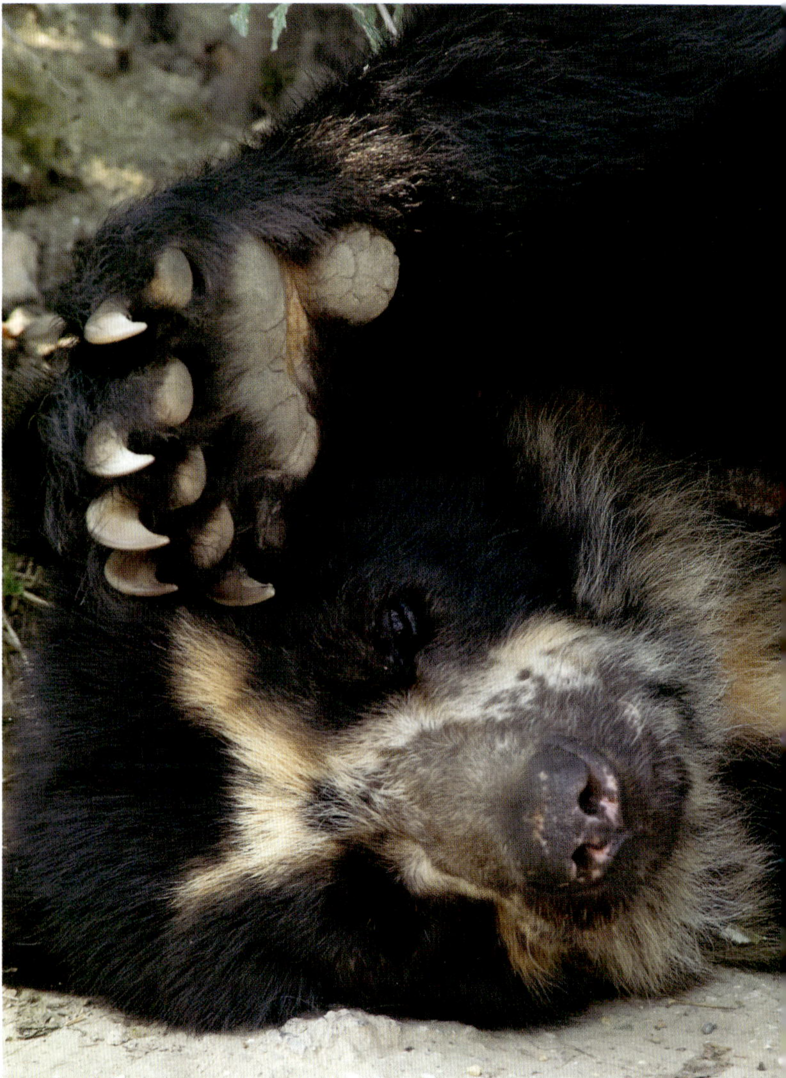

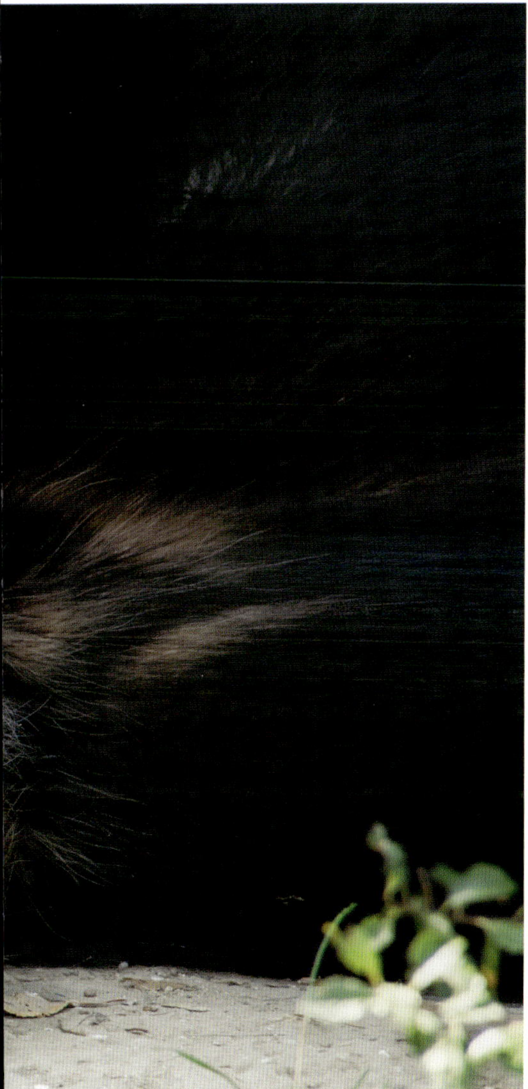

Spectacled bear
The gallbladder of the spectacled bear is a ingredient in traditional Chinese medicines, and despite being proven time and again to be useless, these concoctions remain in demand. As a result, poachers are still on the hunt for these rare bears.

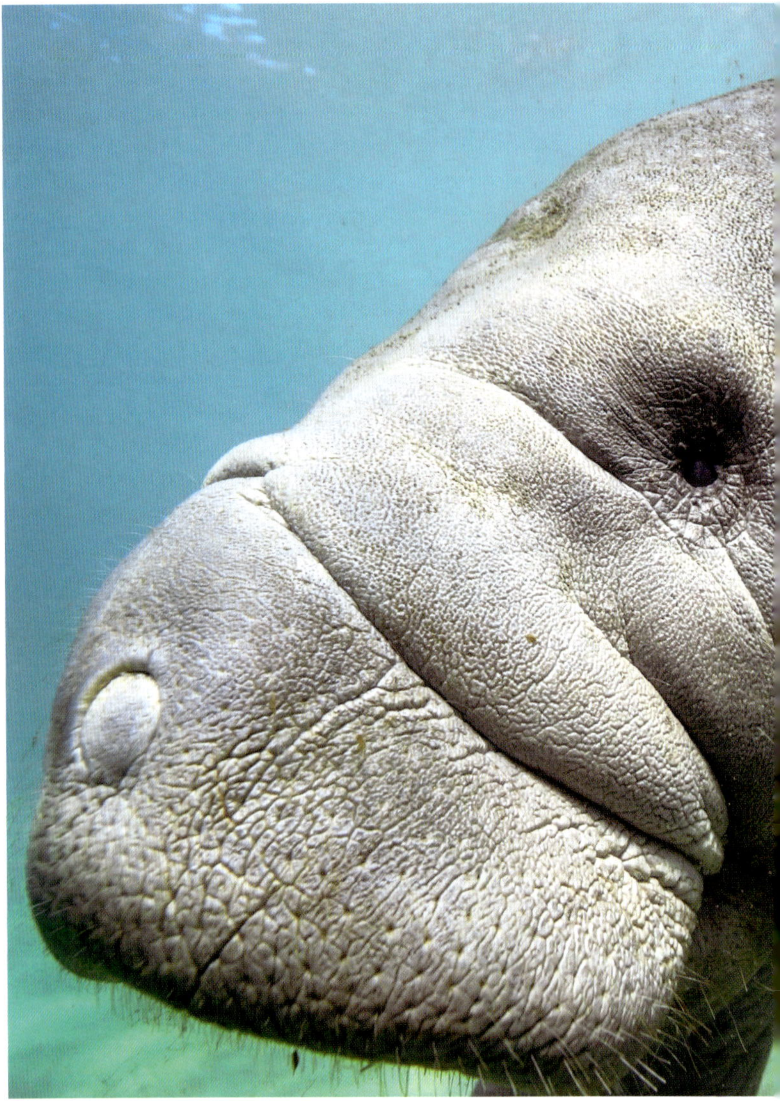

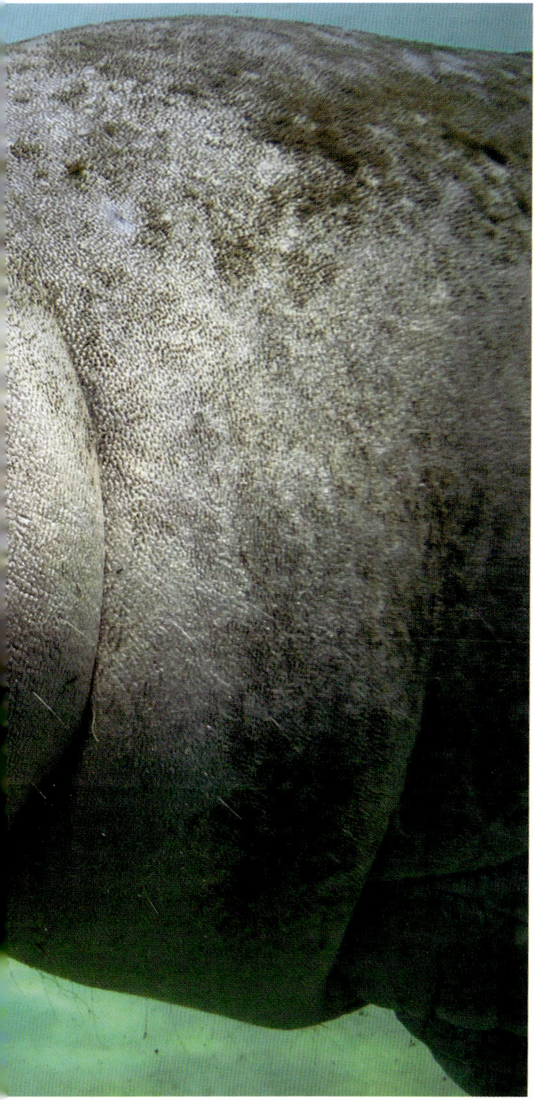

West Indian manatee
One of the three extant species of sea cow – all endangered – this species grazes on sea grasses along the Atlantic coasts of tropical America. They face many threats – the destruction of their feeding sites, poisonous algal blooms caused by pollution, and injury from watercraft that frequent the same area.

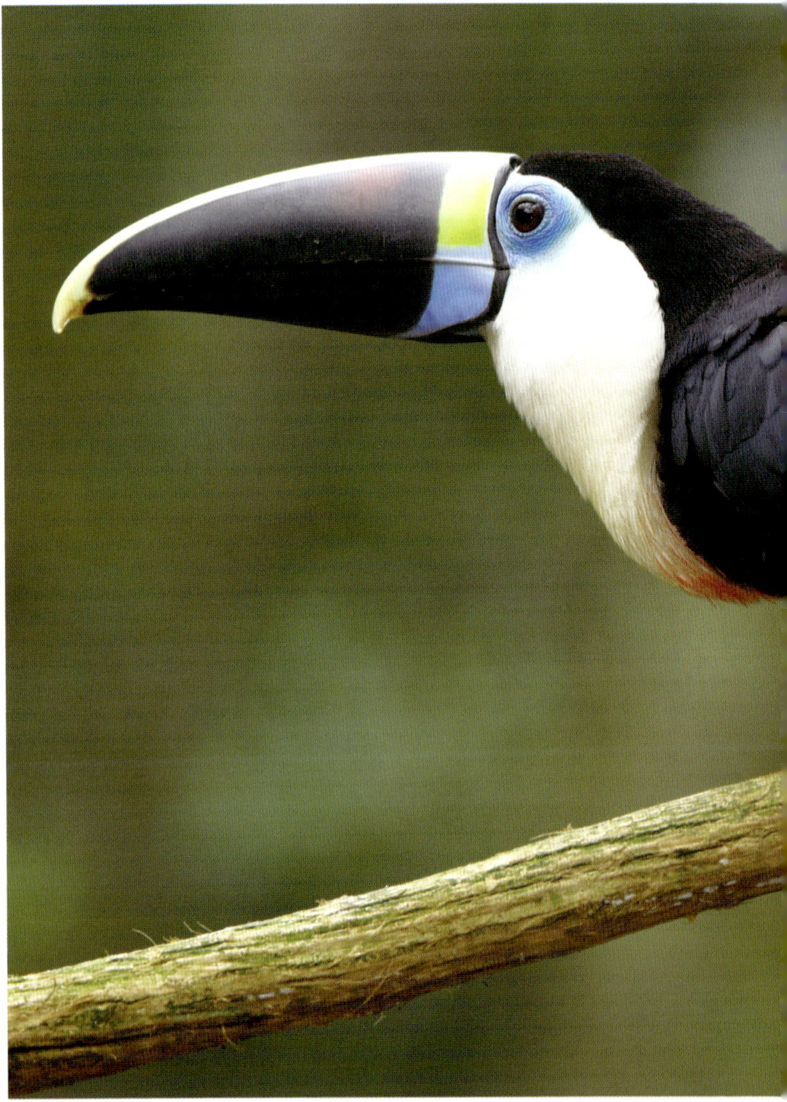

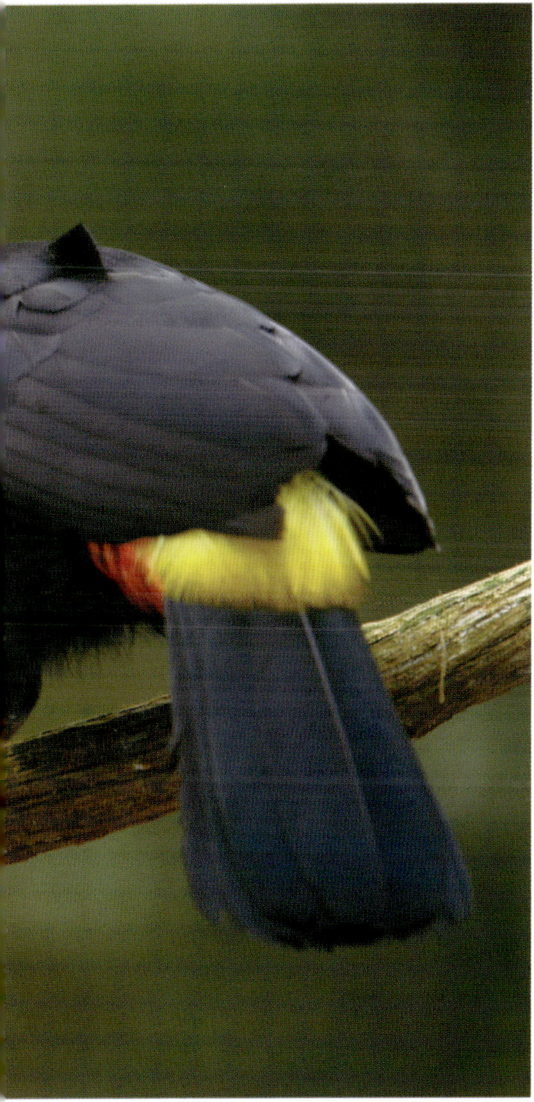

White-throated toucan

Toucans are amazing-looking birds, and this Amazon species is classed as vulnerable because it is being systematically collected from the wild for the pet trade.

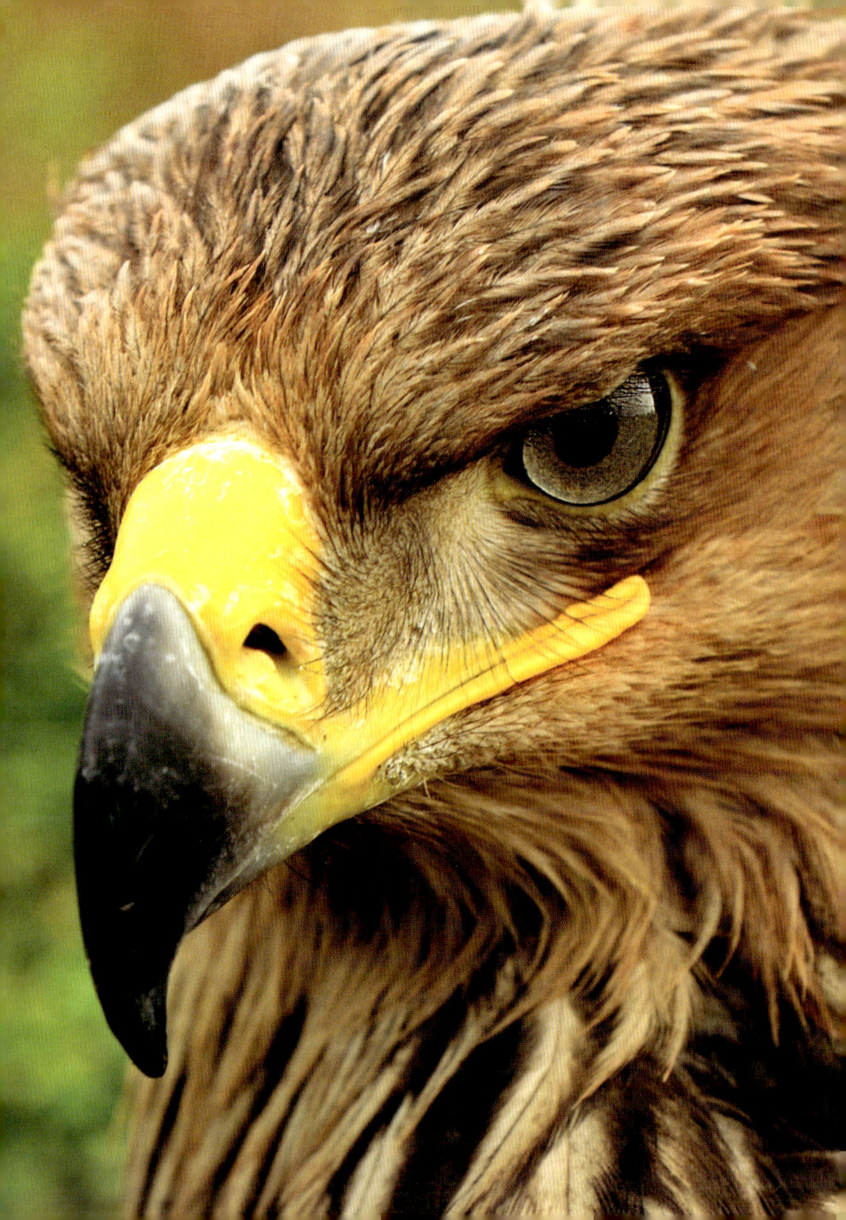

Europe

The human history of Europe is characterised by habitat changes as forests and wetlands were drained to make productive farmland and provide room of cities. This process long precedes the Industrial Revolution, but by the time Europe – particularly the north-western region – began to industrialise in the 1760s, the ecology of the continent changed forever. Today, many places, such as Britain and the Low Countries, are largely devoid of true wilderness. Few wild mammals survive and most of the wildlife to be seen are birds that do well in a countryside that has been transformed into fields.

Decades of pollution and smoke, although now largely cleaned up, led to the collapse of many ecosystems, and what remains is a denuded version of a once verdant land. Nevertheless, European countries have worked hard to preserve what natural habitats they have left, and several species, sich as the Iberian lynx and wisent, or European bison, have been saved from extinction. The value of keystone species such as beavers and wolves is now better recognised, and they are being reintroduced in many places. In the decades to come, they will help recreate the ancient landscapes and provide new spaces for wildlife to thrive.

OPPOSITE:

Steppe eagle

Spread across the grasslands of Asia, this mighty eagle clings on in the southeast corner of Europe in the Caucuses.

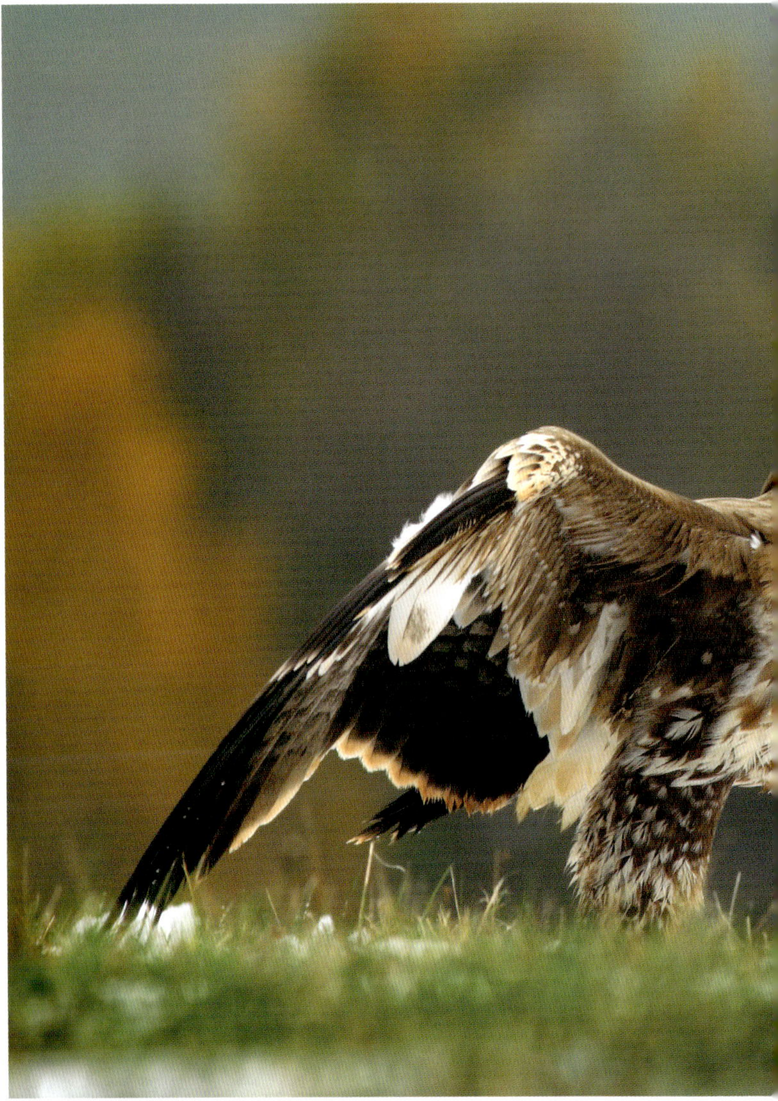

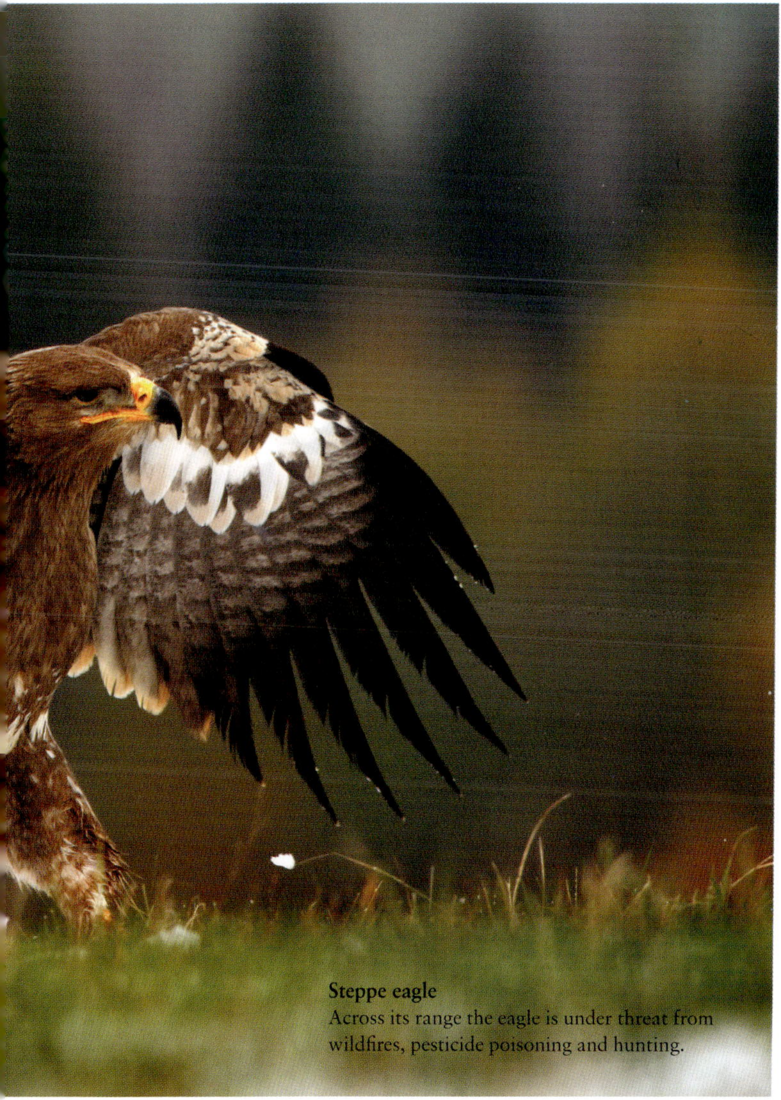

Steppe eagle
Across its range the eagle is under threat from wildfires, pesticide poisoning and hunting.

European sturgeon

The largest river fish in Europe, and one of the biggest in the world, this fish is a source of caviar, which are its eggs. The eggs are carried internally, and the fish must be killed to extract them. As a result, the adult population of this and all sturgeon species has plunged to almost nothing and breeding is very limited.

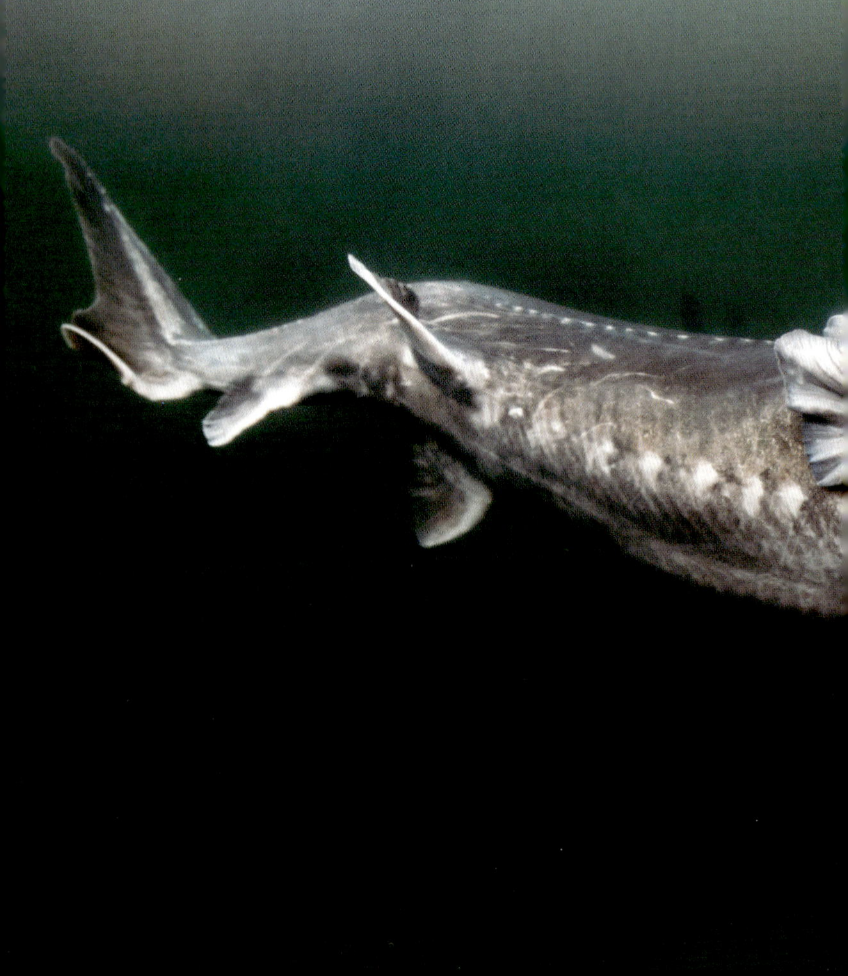

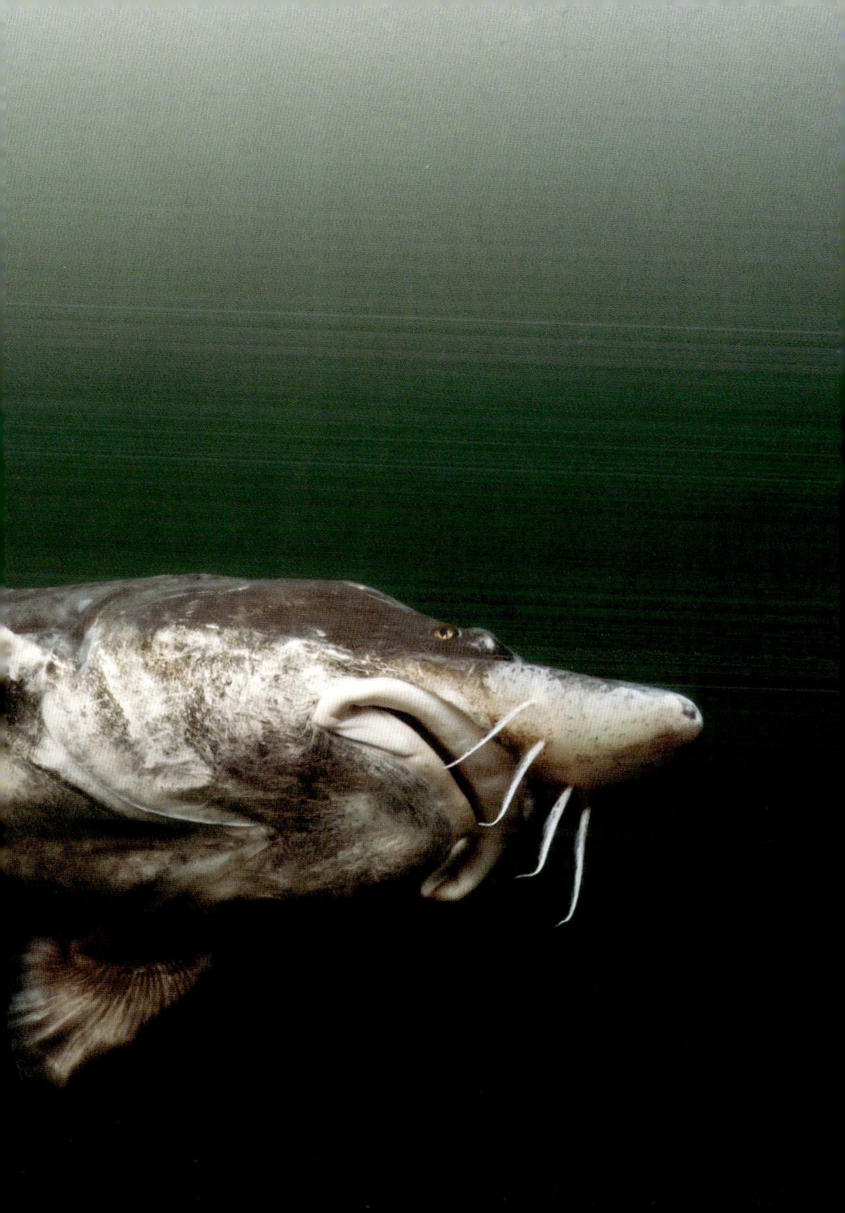

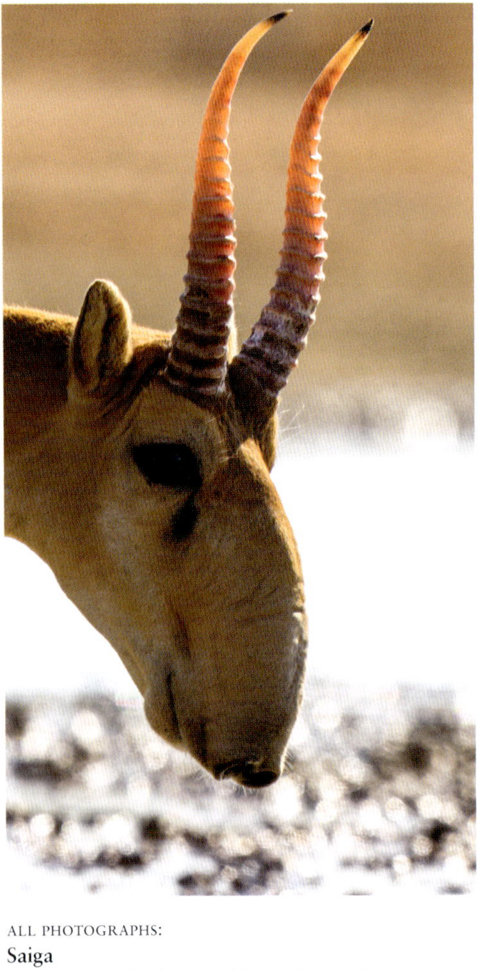

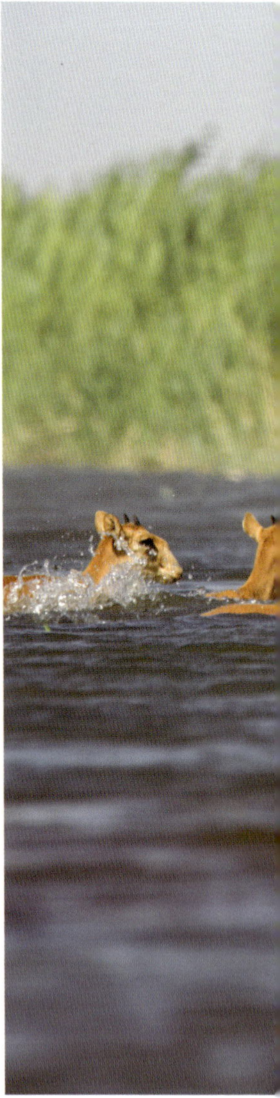

ALL PHOTOGRAPHS:

Saiga

Once vast herds of this odd hoofed animal paraded across the plains of Eastern Europe and Central Asia. Hunting and agriculture have destroyed the population. However, after a long decline, saiga herds appear to be growing again in Kazakhstan.

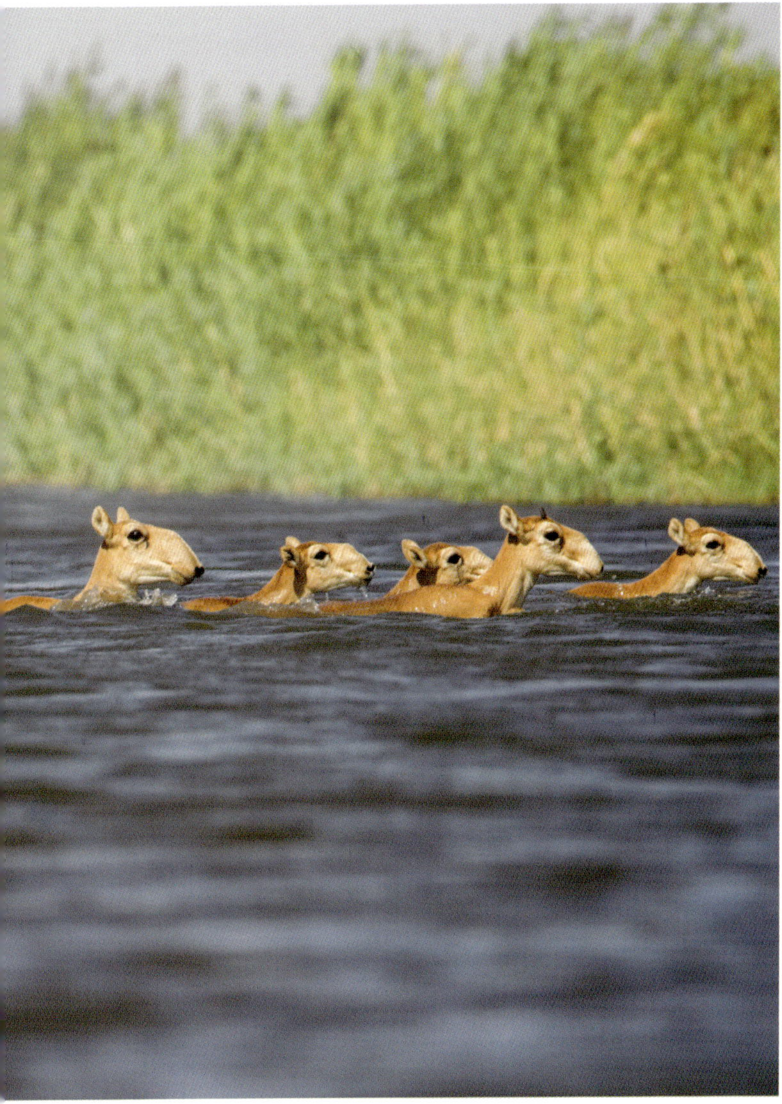

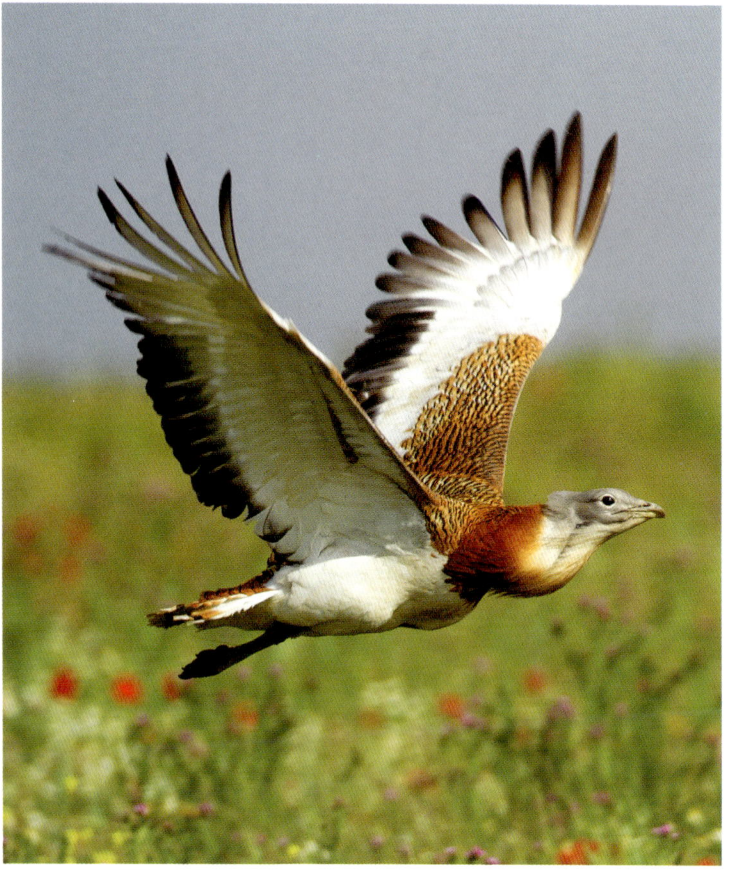

ABOVE:

Great bustard

One of the heaviest flying birds living anywhere in the world, this omnivorous bird once thrived across Eurasia. Today, two-thirds of the population survives in Spain and Portugal, with deforestation, pesticide poisoning and attacks from feral dogs decimating the birds elsewhere.

BELOW AND BOTTOM:

Wisent

Also known as the European bison, this big herd animal has suffered as the continent's woodlands and forests have been systematically reduced for many centuries.

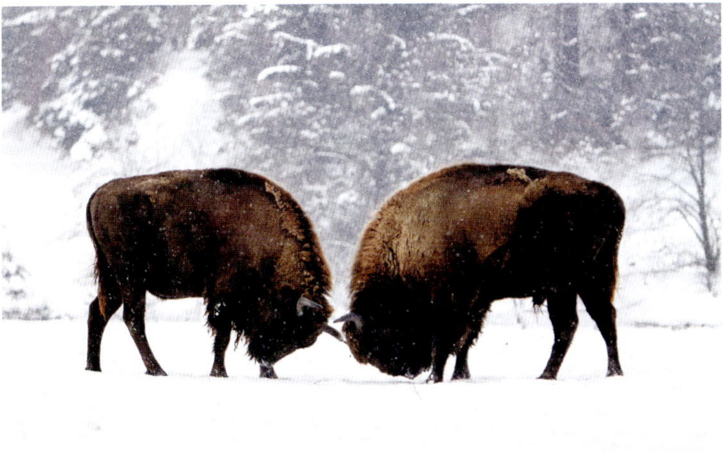

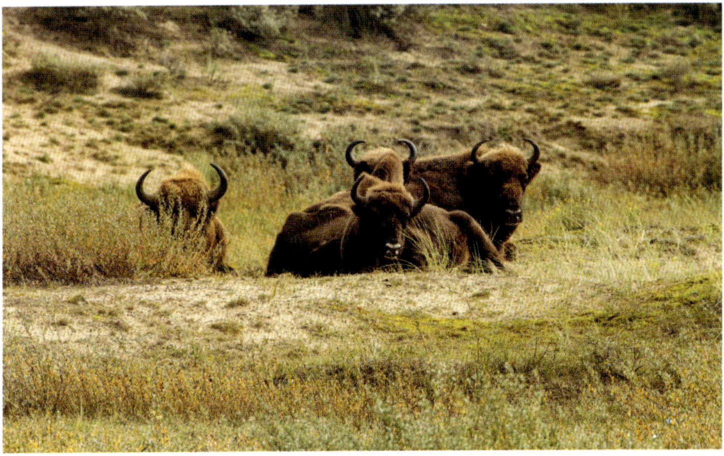

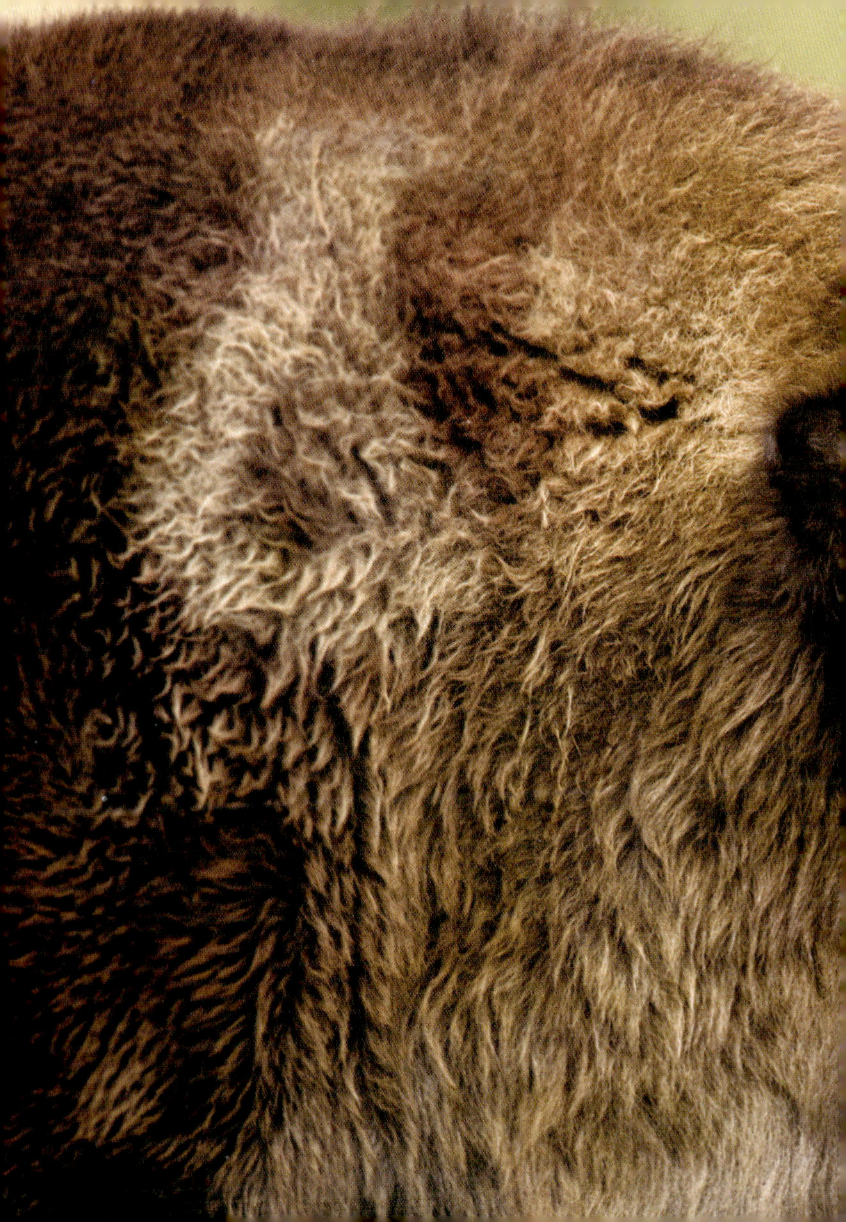

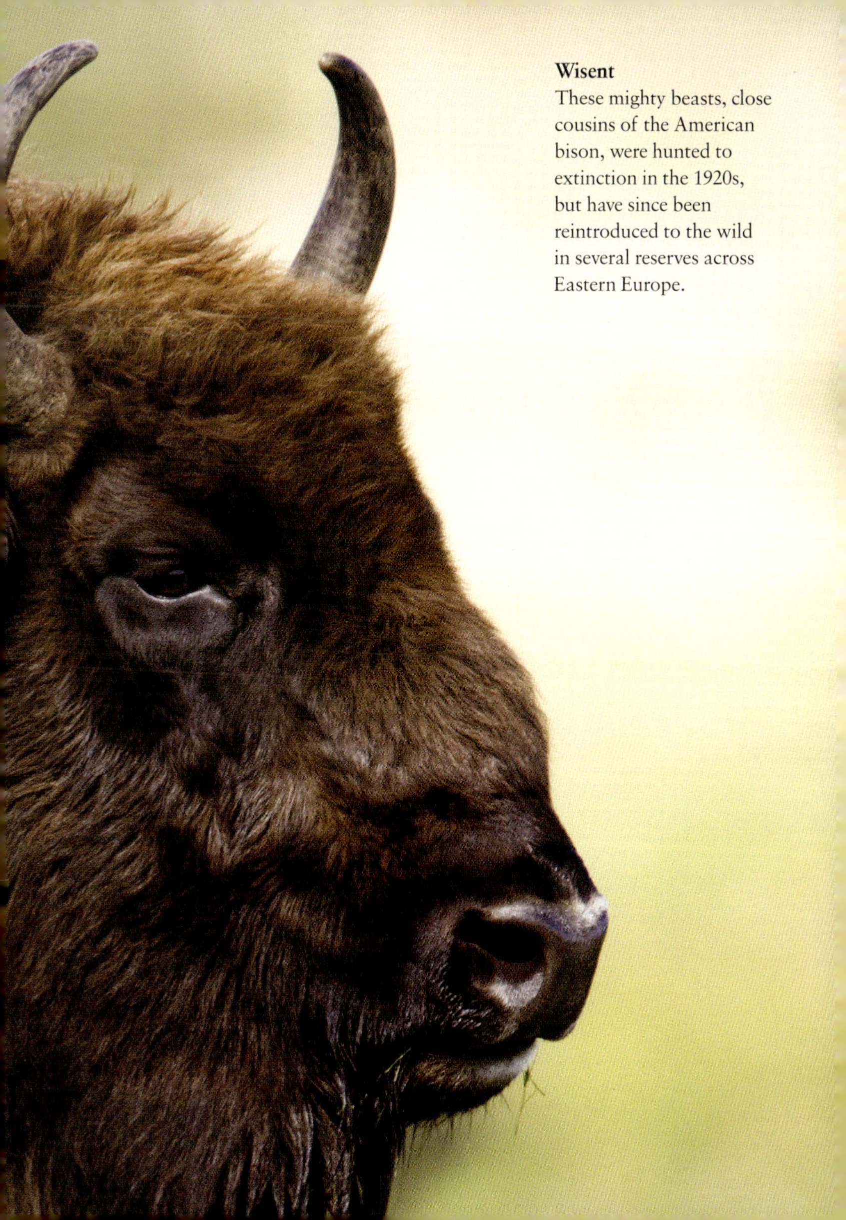

Wisent

These mighty beasts, close cousins of the American bison, were hunted to extinction in the 1920s, but have since been reintroduced to the wild in several reserves across Eastern Europe.

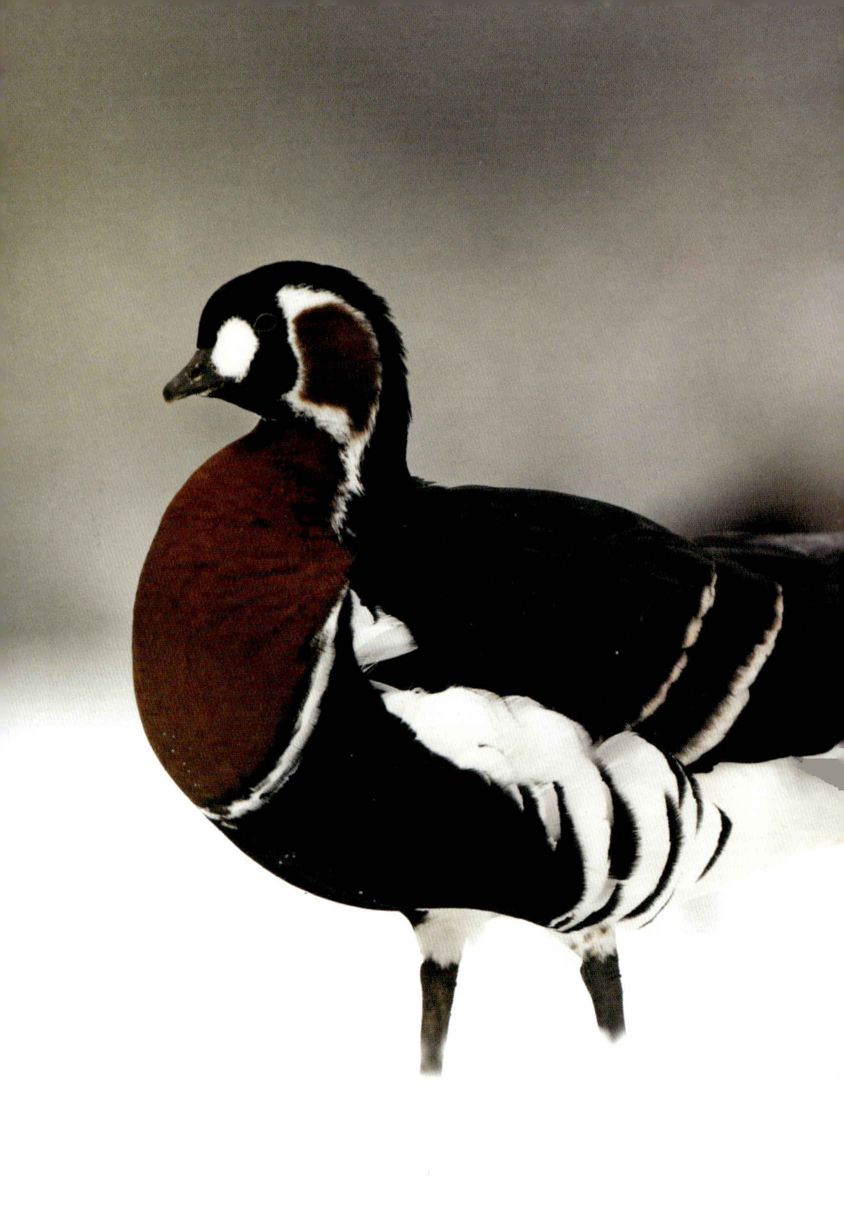

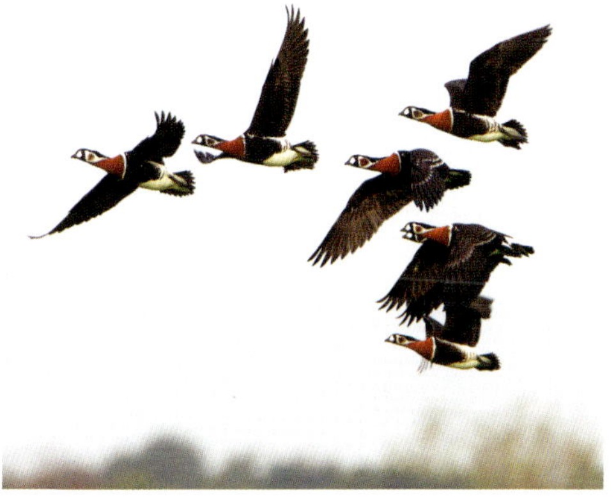

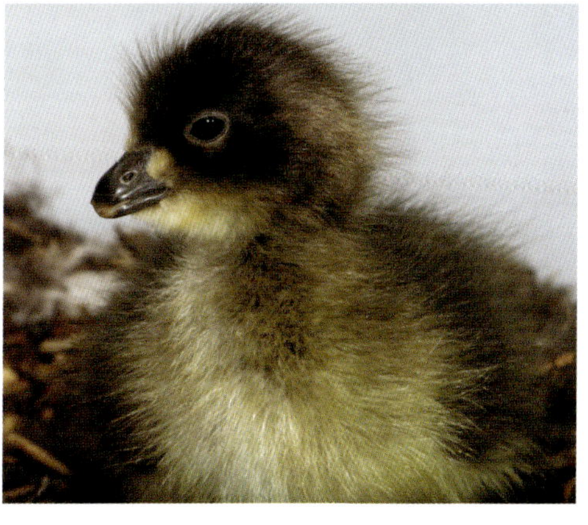

ALL PHOTOGRAPHS:

Red-breasted goose

This small but elaborately coloured goose migrates to the Arctic in western Siberia in the summer and returns to wintering grounds around the western and northern coasts of the Black Sea. The global population has recently dropped to 70,000, and the decline continues.

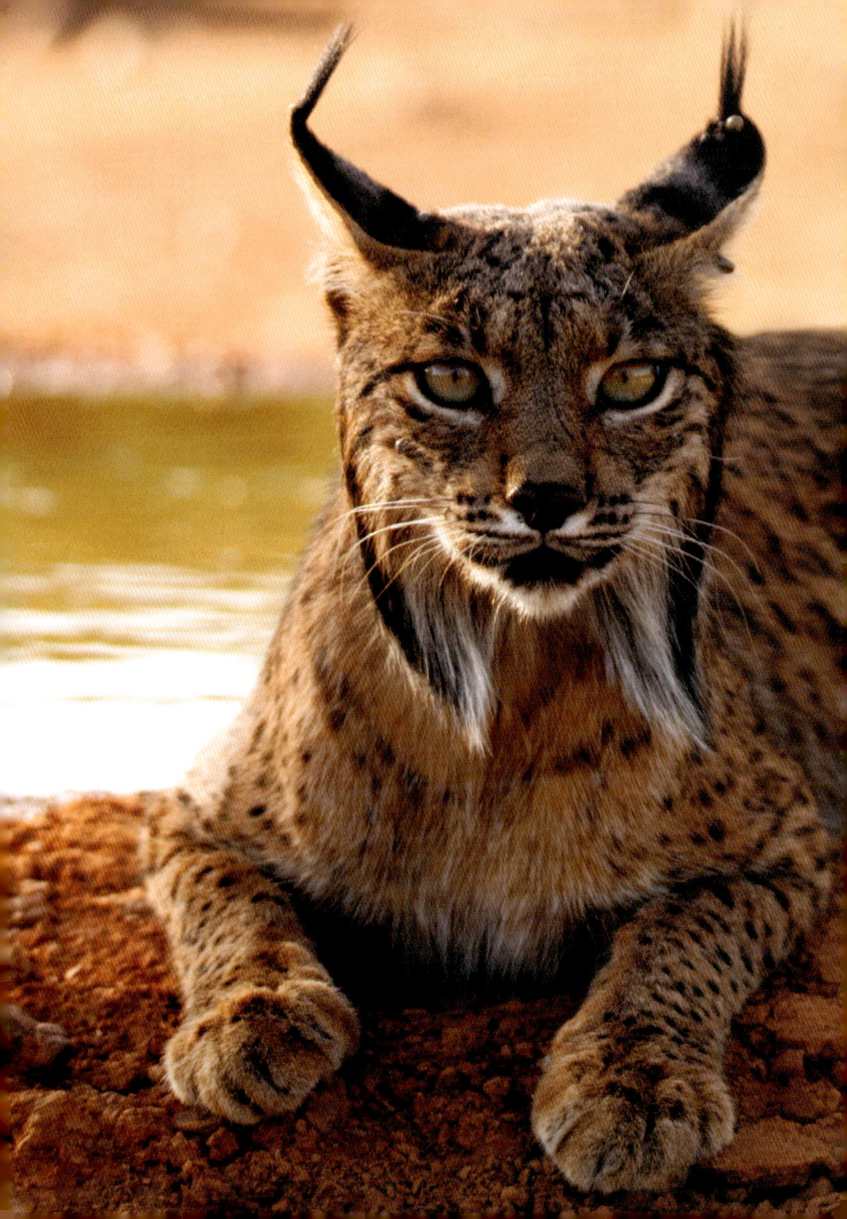

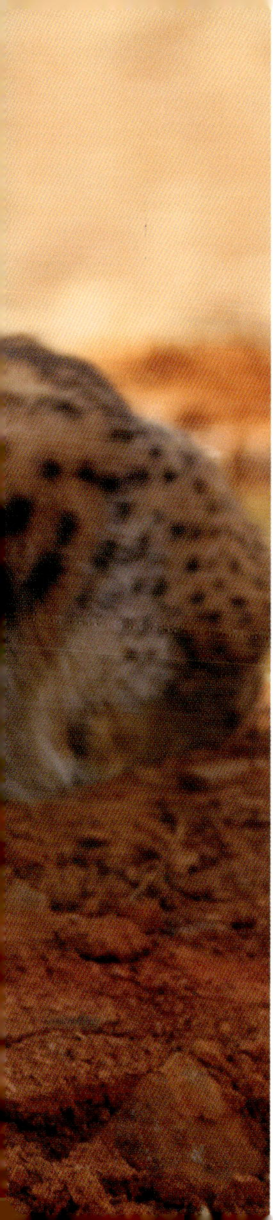

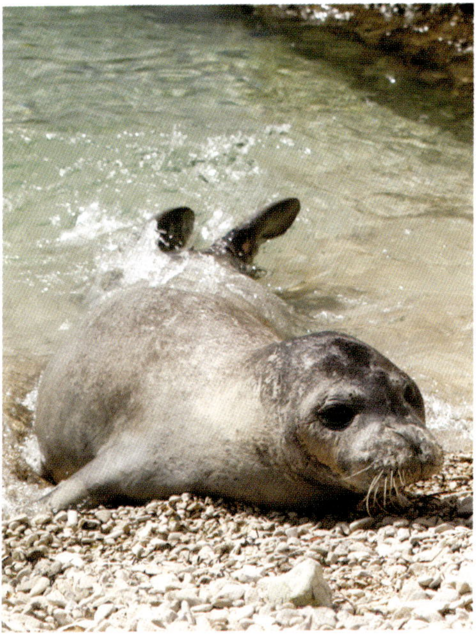

LEFT:

Iberian lynx

By the early 1990s this small hunting cat's population was reduced to just a few dozen in the wild. The cats are specialist hunters of rabbits, and the lynx's demise was due to a crash in the rabbit population in the area. A captive breeding and reintroduction programme has brought the species back from the brink.

ABOVE:

Mediterranean monk seal

It is estimated that just 400 adult seals survive in the wild, mostly in the Aegean Sea. However, conservationists have succeeded in turning the tide, and the monk seal population is rising again.

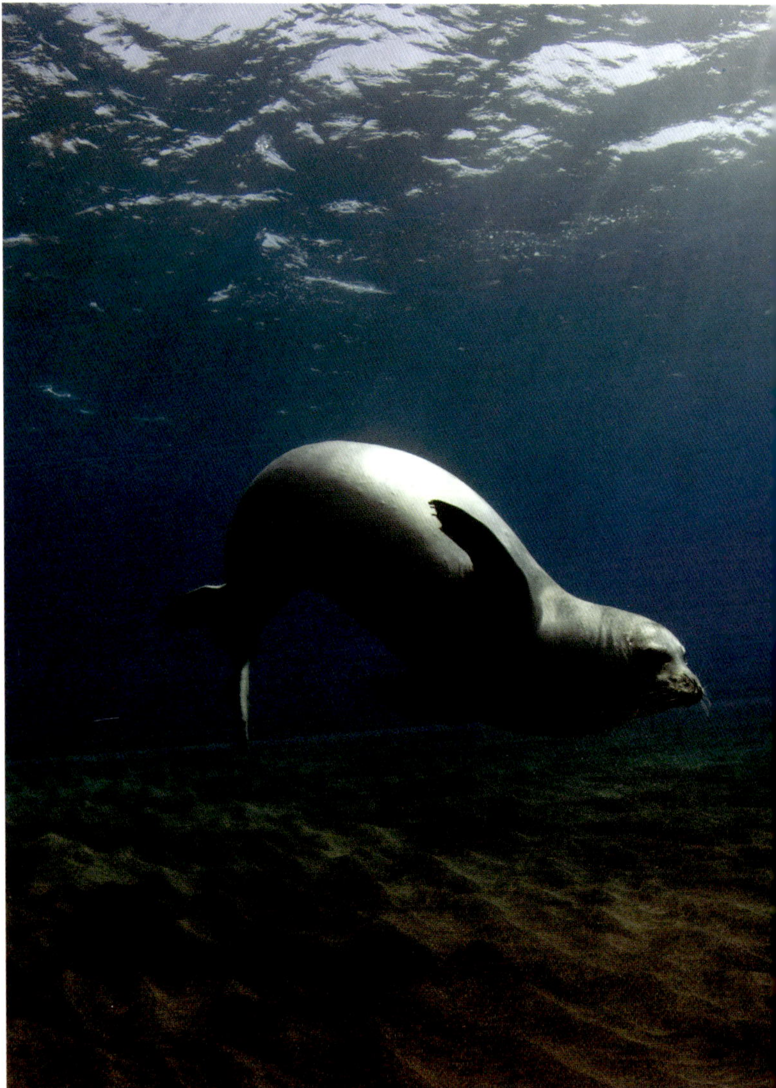

ALL PHOTOGRAPHS:

Mediterranean monk seal

This small seal has been badly impacted by the expansion of shipping routes, pollution and the increase in fishing. Additionally, development along coastal areas has reduced the number of viable seal breeding sites.

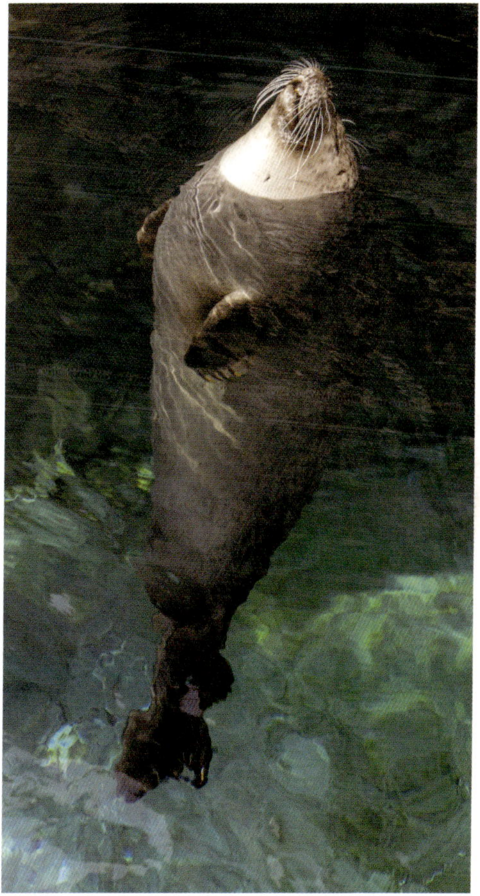

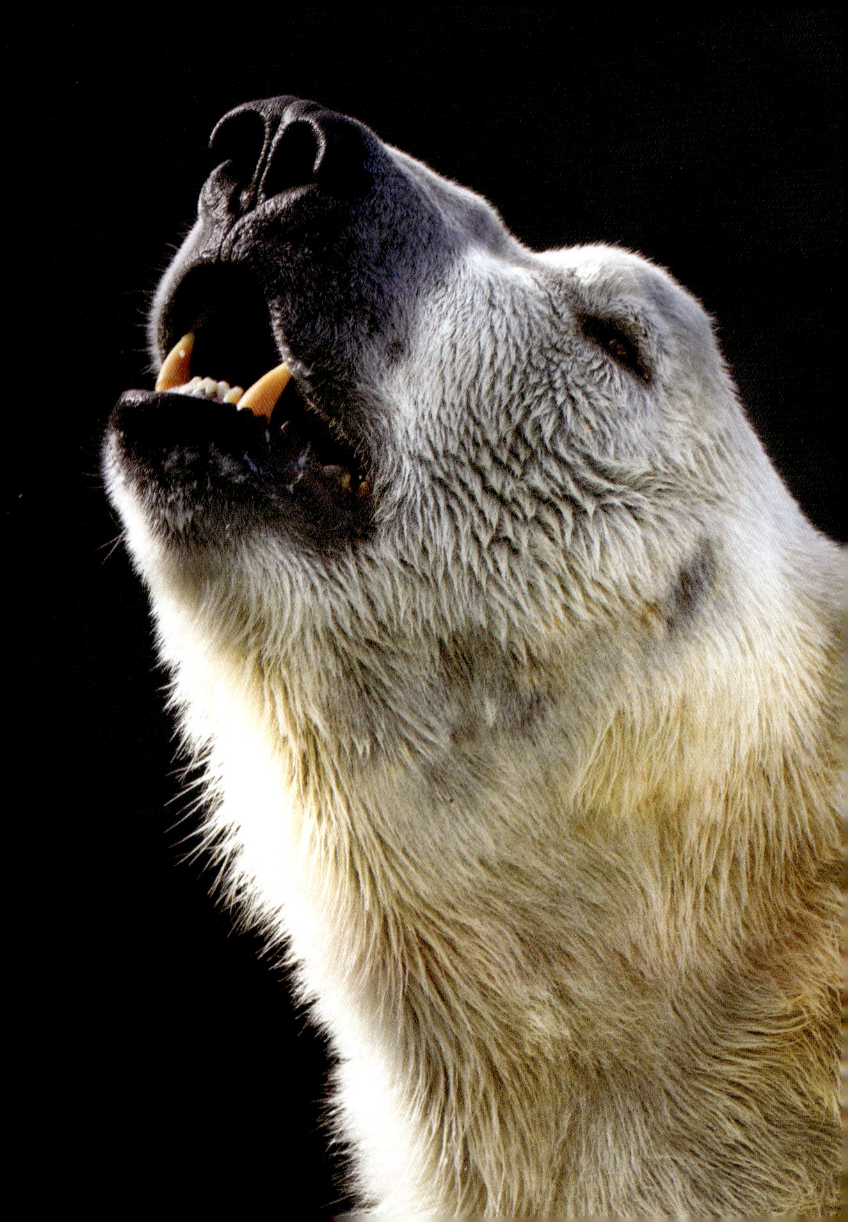

Polar Regions and Oceans

The polar regions are the driest and coldest places on Earth – there is no liquid water to drink when it is always freezing outside. As a result, only a very few of the hardiest land animals survive in the Arctic and Antarctic, where it is dark for many months in winter, and all the opportunities to feed and breed are confined to a few short months in summer. The Arctic is warming at twice the rate of other parts of the planet due to climate change, and so those animals supremely adapted to the rhythm of the ice are struggling to adapt to a new reality where the seas and land are no longer frozen. Only time will tell if the Arctic, named after its most imposing resident, the polar bear ('bear' translates to 'arctos' in Greek) will remain a place where these mighty hunters can survive.

The oceans are being impacted by climate change as well, and even small increases in water temperature can reduce the quantity of oxygen dissolved in it, thus compressing most sea life into a shrinking band of water beneath the surface. The direct activities of humans, such as fishing and pollution, are also having an impact on the balance of marine ecosystems. More that 5,000 ocean species are listed as endangered, and people are realising that our vast seas are not an endless resource.

OPPOSITE:

Polar bear

Adapted to life on the Arctic ice in summer, the polar bear is in serious peril as climate change reduces it habitat.

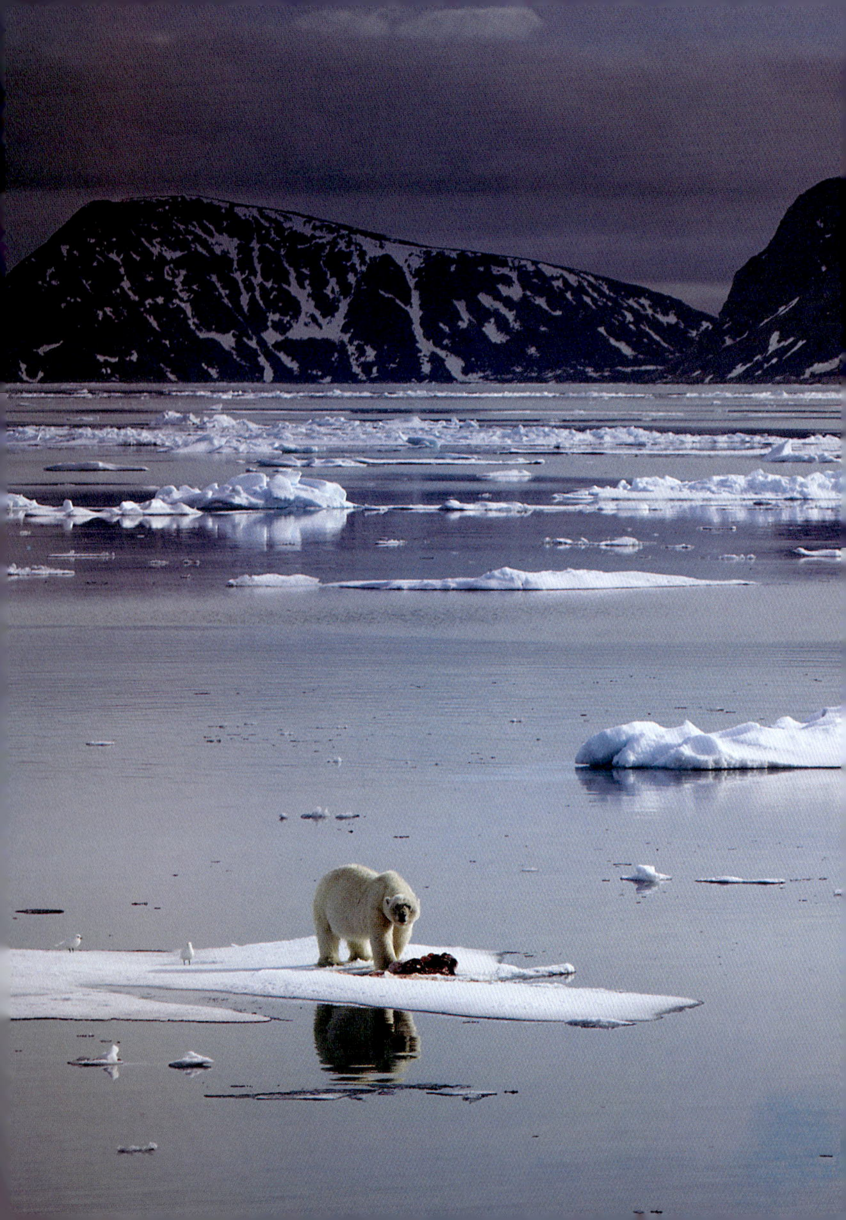

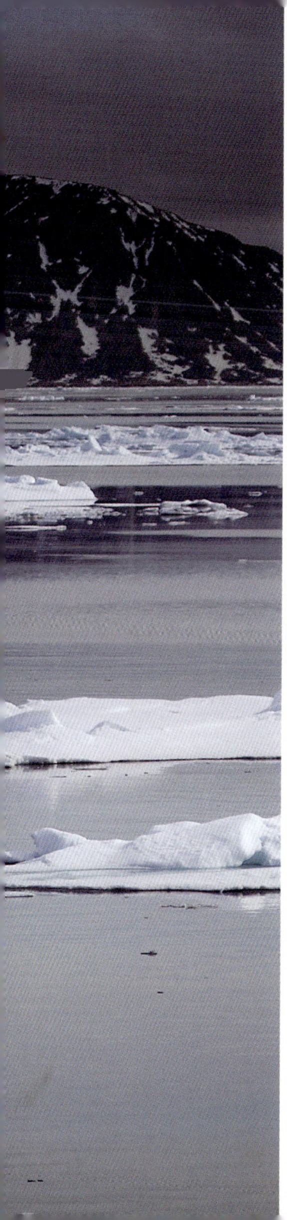

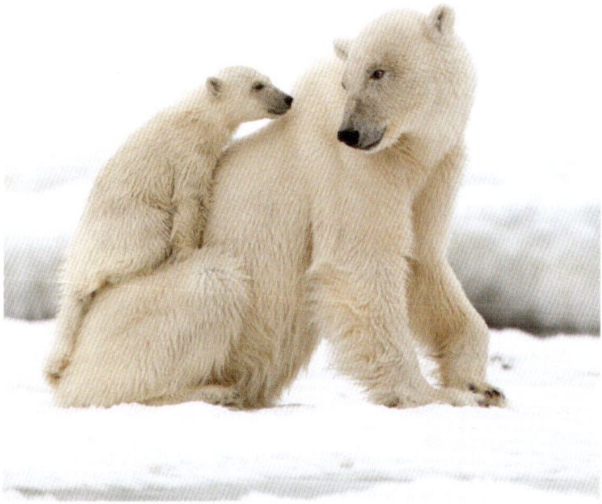

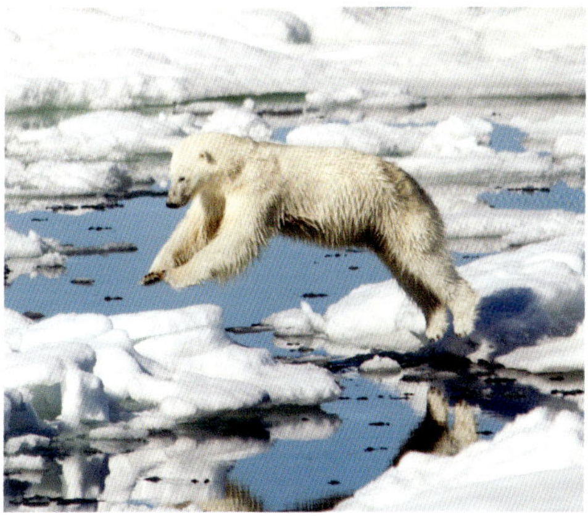

ALL PHOTOGRAPHS:

Polar bear

The white bear is well able to cross open water, but it needs solid ice to target seals, its main food source. As the ice cover shrinks, hungry bears are forced south on to land to find enough food before it hibernates. This means bears are meeting humans more frequently.

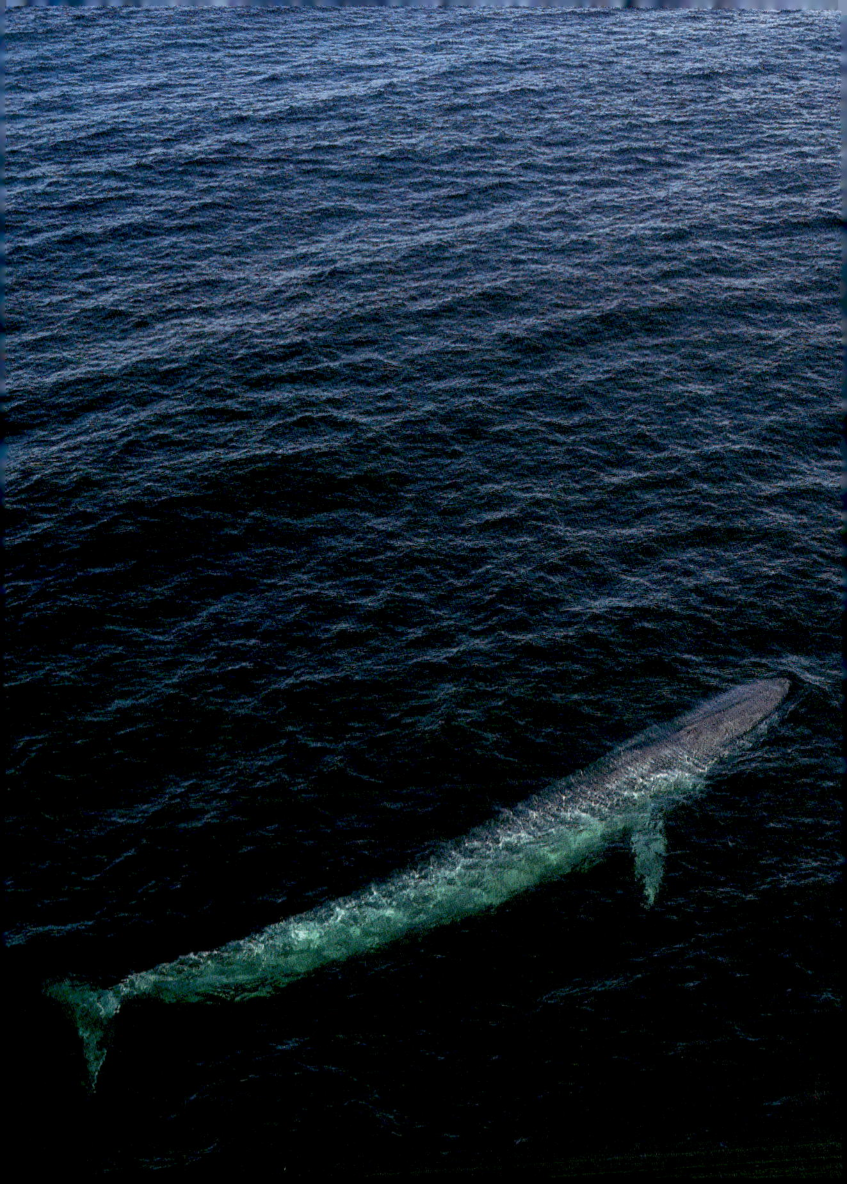

Blue whale

The mightiest creature that has ever lived on Earth, the blue whale was close to extinction as many thousands were killed each year until the 1960s. Today numbers are rising – but slowly.

Blue whale

The blue whale cruises the world's oceans, eating vast mouthfuls of planktonic krill. An upper estimates of the blue whale population is 15,000.

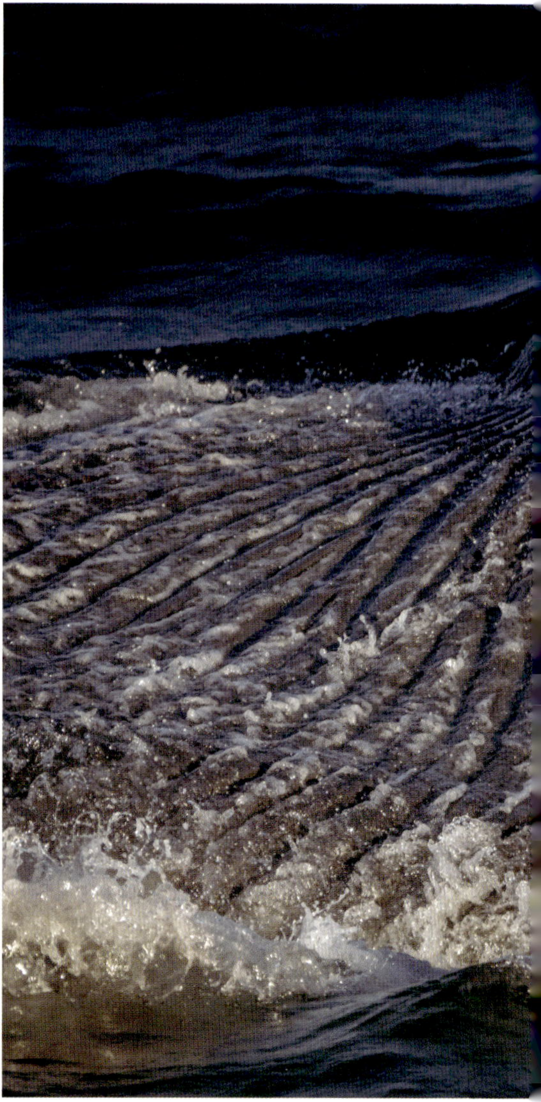

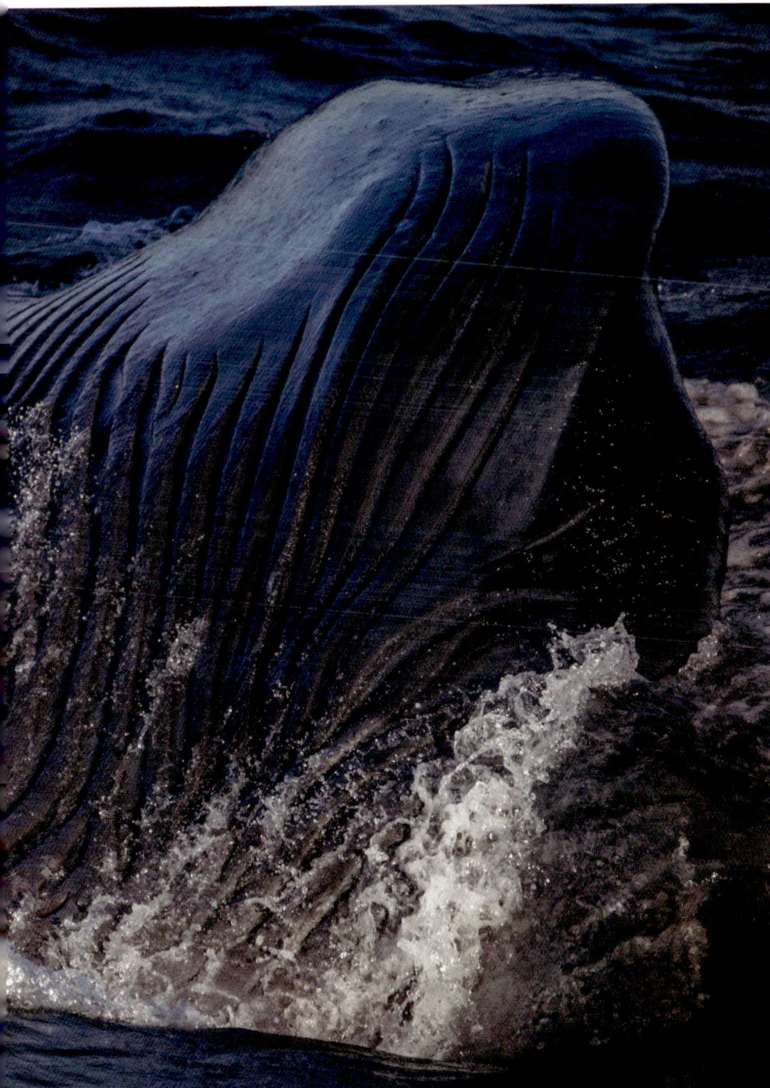

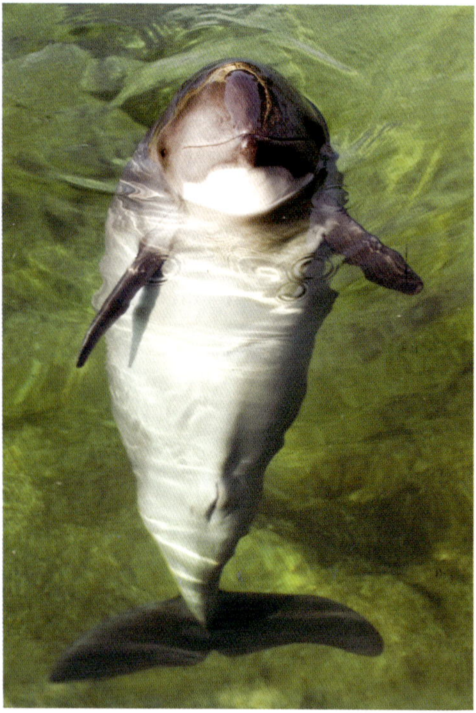

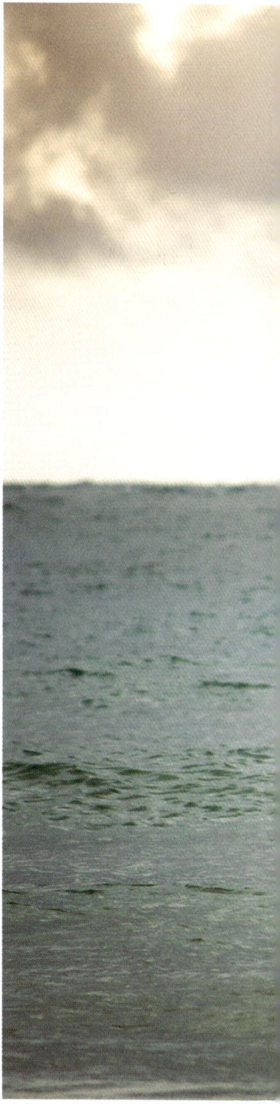

ABOVE:

Harbour porpoise

Porpoises are small relatives of the dolphins, easily spotted by the short, blunt snout. There are seven species and most of them are endangered. The harbour porpoise lives across the Northern hemisphere and is threatened, mainly by fishing nets.

RIGHT:

Hector's dolphin

Found only around the coasts of New Zealand, this species of dolphin is under threat from accidents with boats and fishing nets. The population has dropped by two-thirds in the last 40 years.

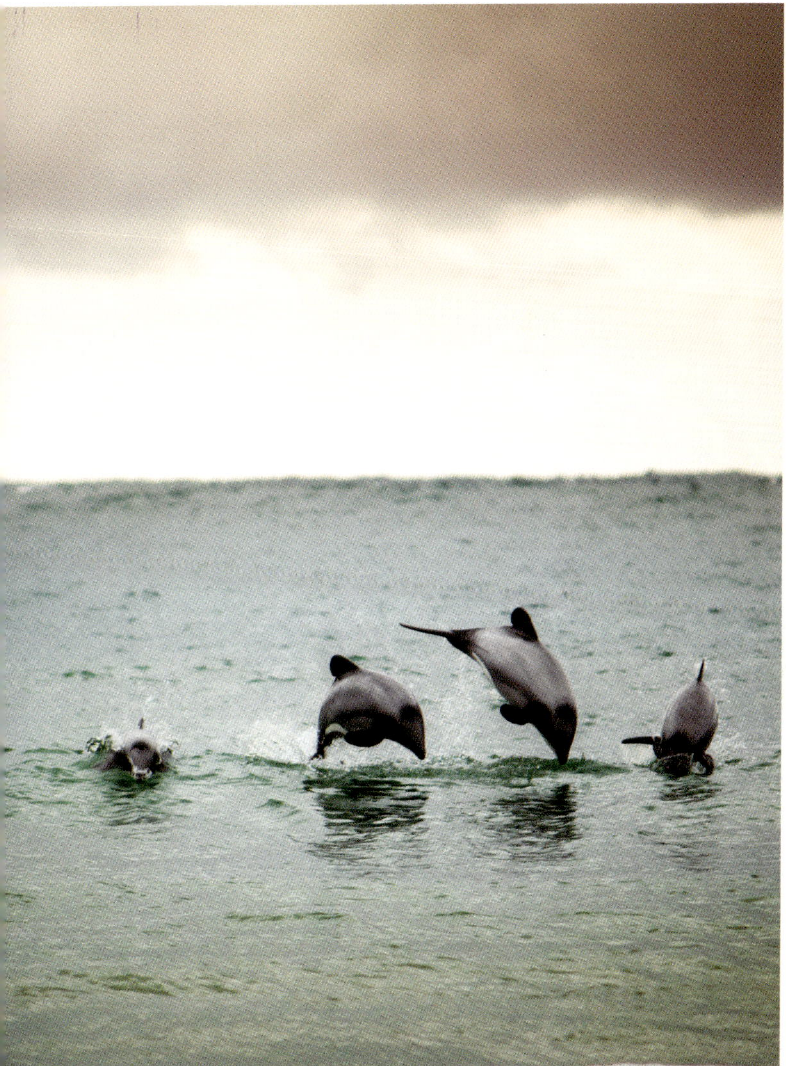

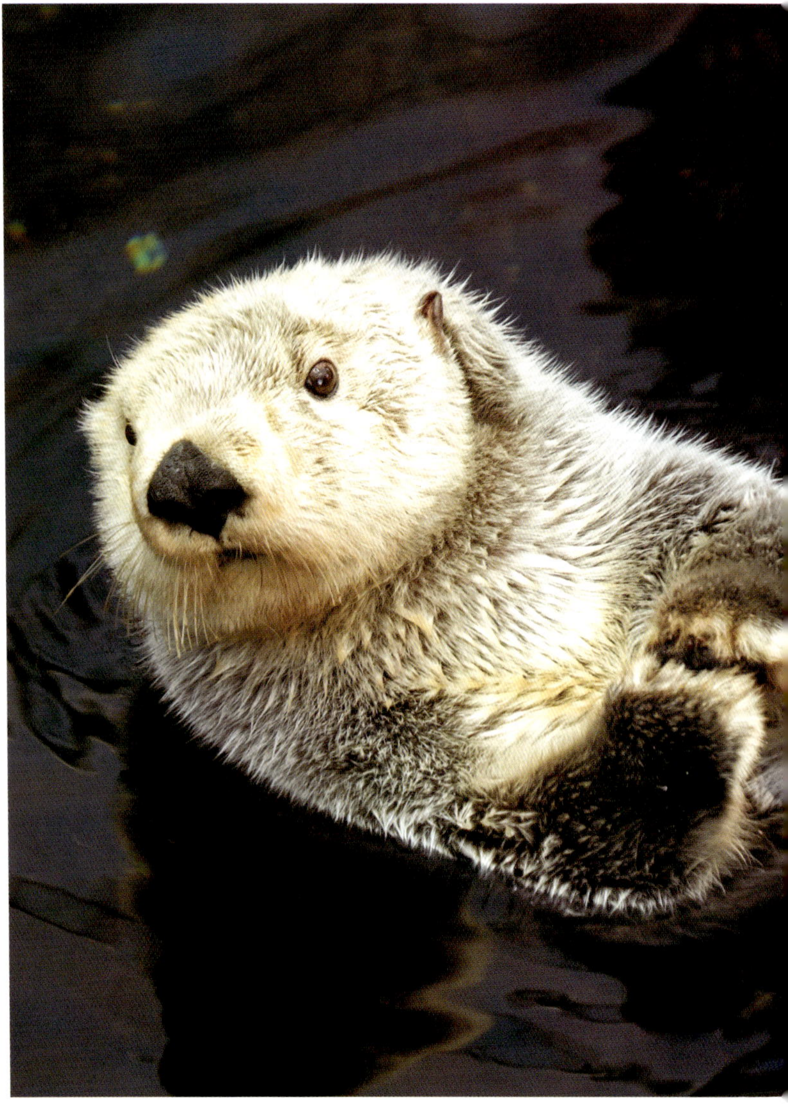

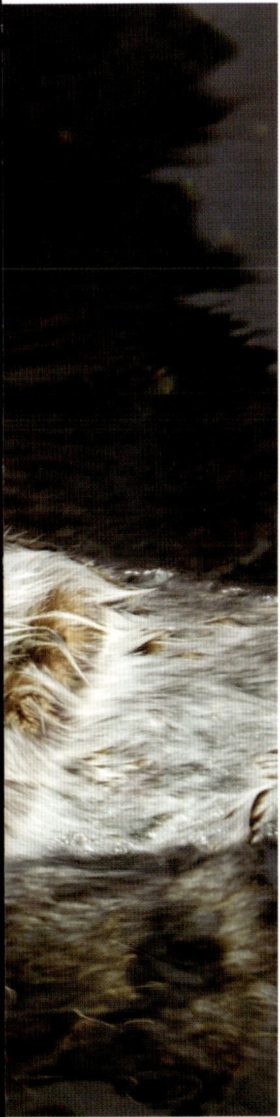

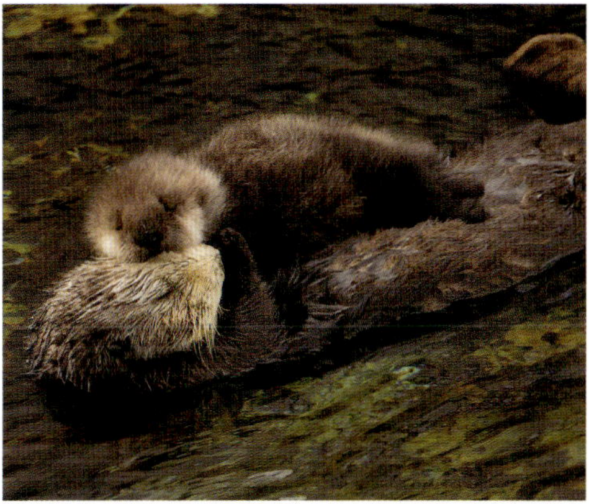

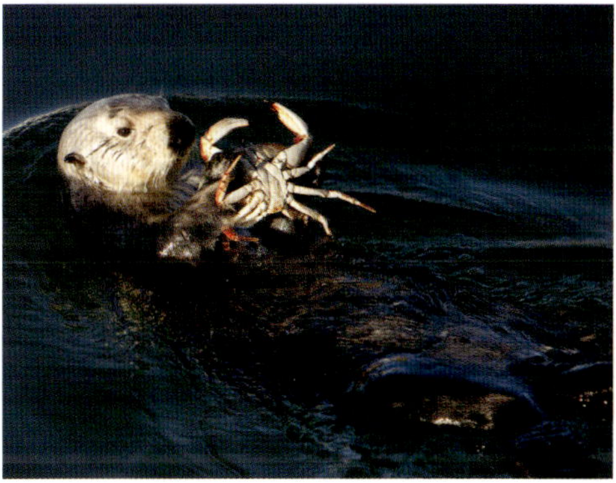

ALL PHOTOGRAPHS:

Sea otter

Rightly celebrated as one of the cutest creatures on the planet, the fur trade along the Pacific coast drove them close to extinction at the start of the 20th century. Without enough otters harvesting sea urchins from the seabed, the urchins destroyed the dense kelp forests that grow offshore. Without the forests, ocean waves accelerated coastal erosion. Today, these habitats are protected and the sea otters and ecological balance are returning.

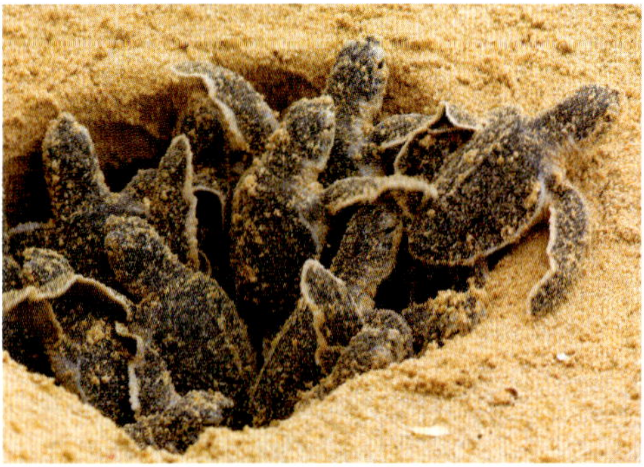

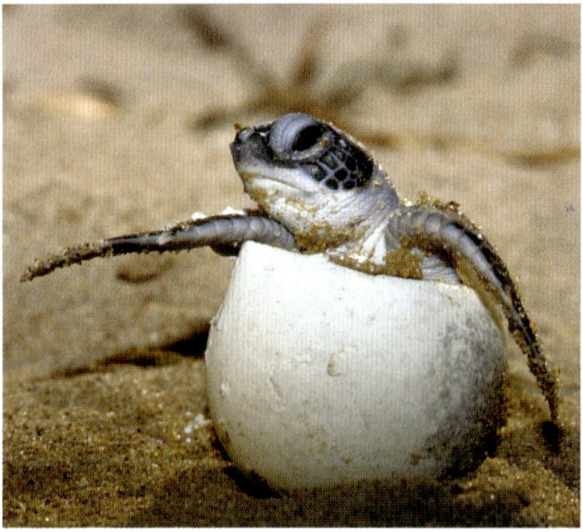

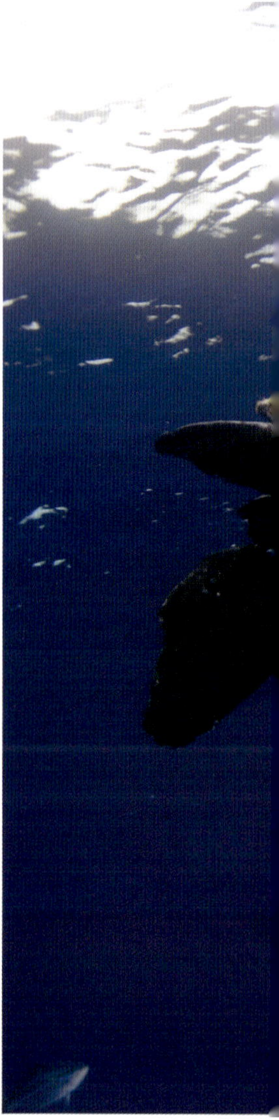

ALL PHOTOGRAPHS:

Leatherback turtle

The largest turtle in the world, characterised by its soft leathery carapaces instead of a hard shell, the leatherback is vulnerable all over the world. Females always return to the same beach to lay eggs, the same one where they hatched themselves. Human intrusion on turtle beaches is reducing breeding opportunities.

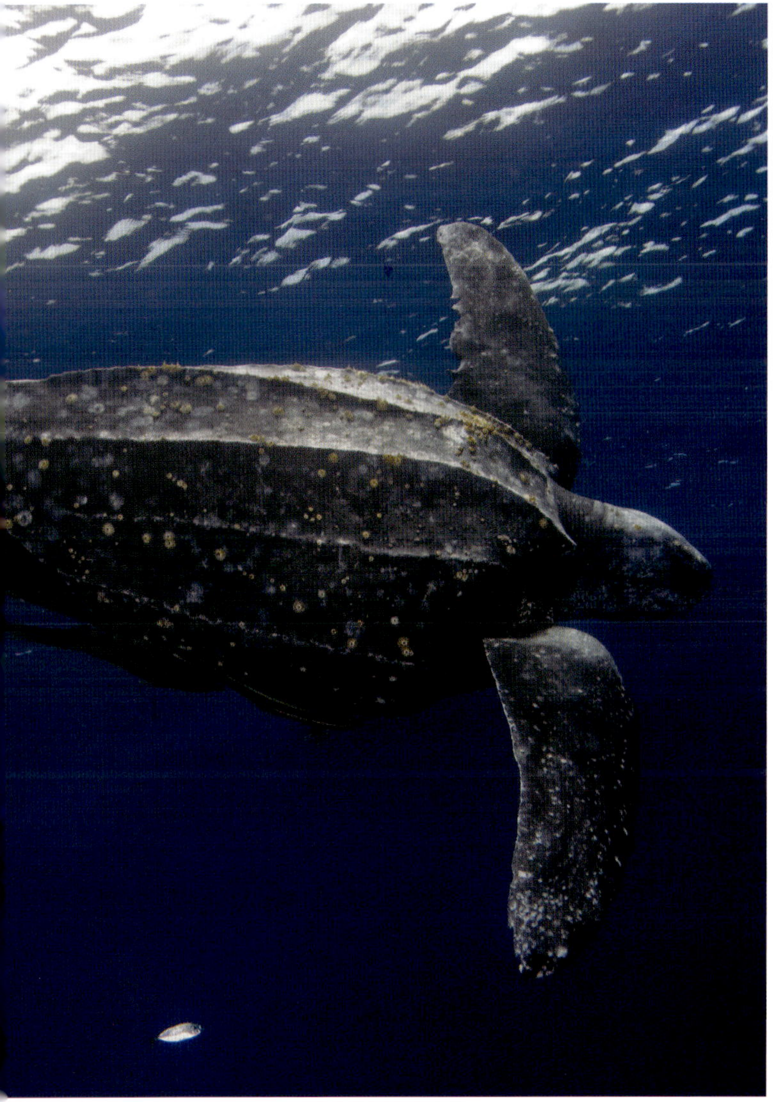

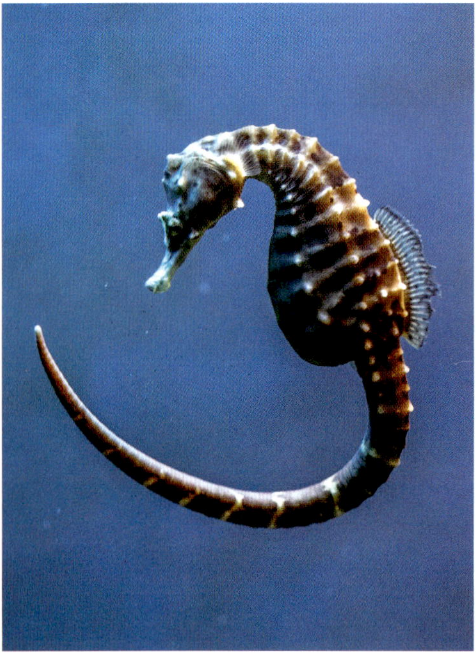

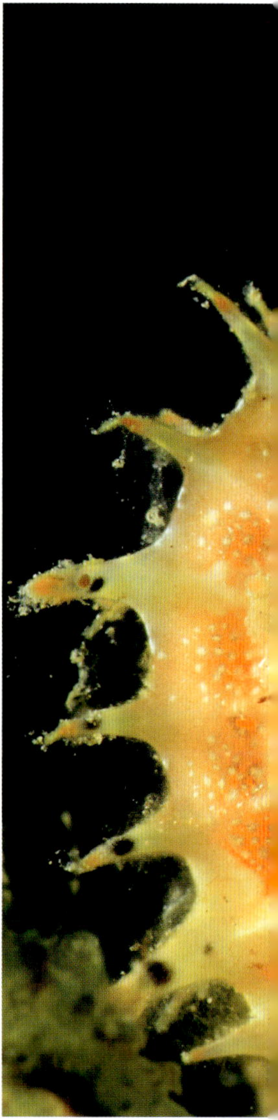

ALL PHOTOGRAPHS:

Seahorse

There are 46 species of seahorse, mostly living in the seaweed beds of shallow seas. They cling to plants with their flexible tails and suck food particles from the water using their equine mouthparts. Being dependant on these fragile habitats means that most, if not all, seahorses are endangered.

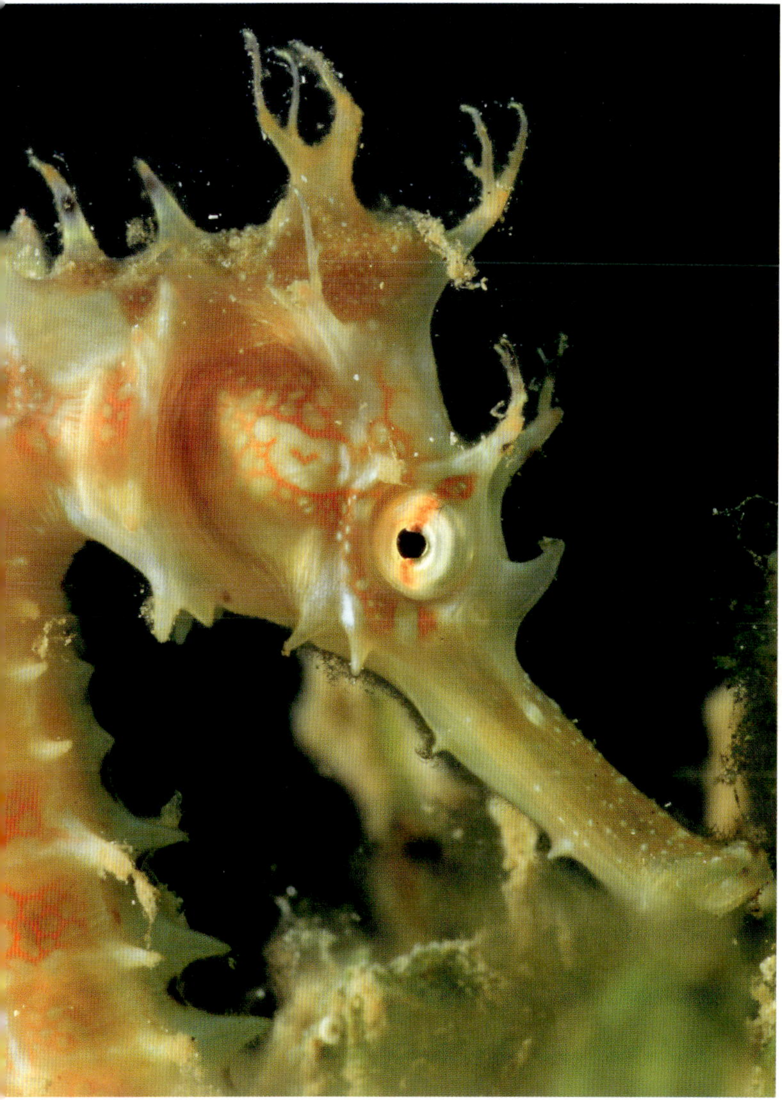

Great white shark

Famed as the biggest and fiercest hunting fish in the ocean, the great white shark is persecuted the world over. It does attack humans and one bite can often be fatal, so great whites may be killed if they enter coastal waters. However, they are more likely to be killed in fishing nets. Little is known about shark numbers but it is assumed that the population is in decline.

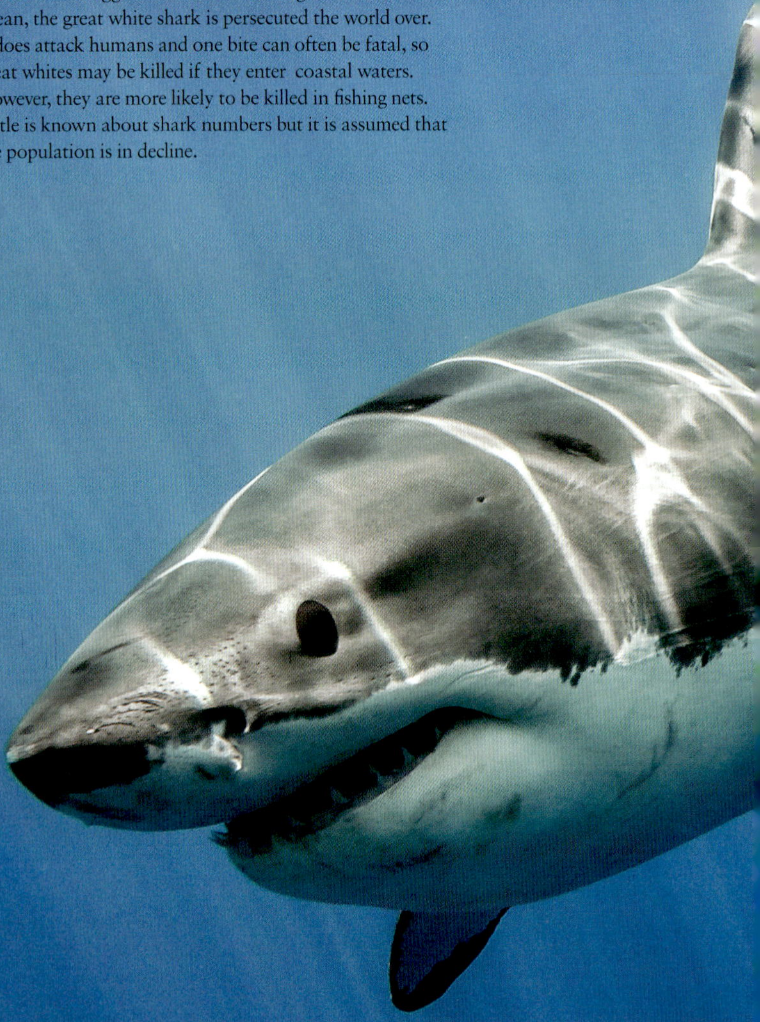

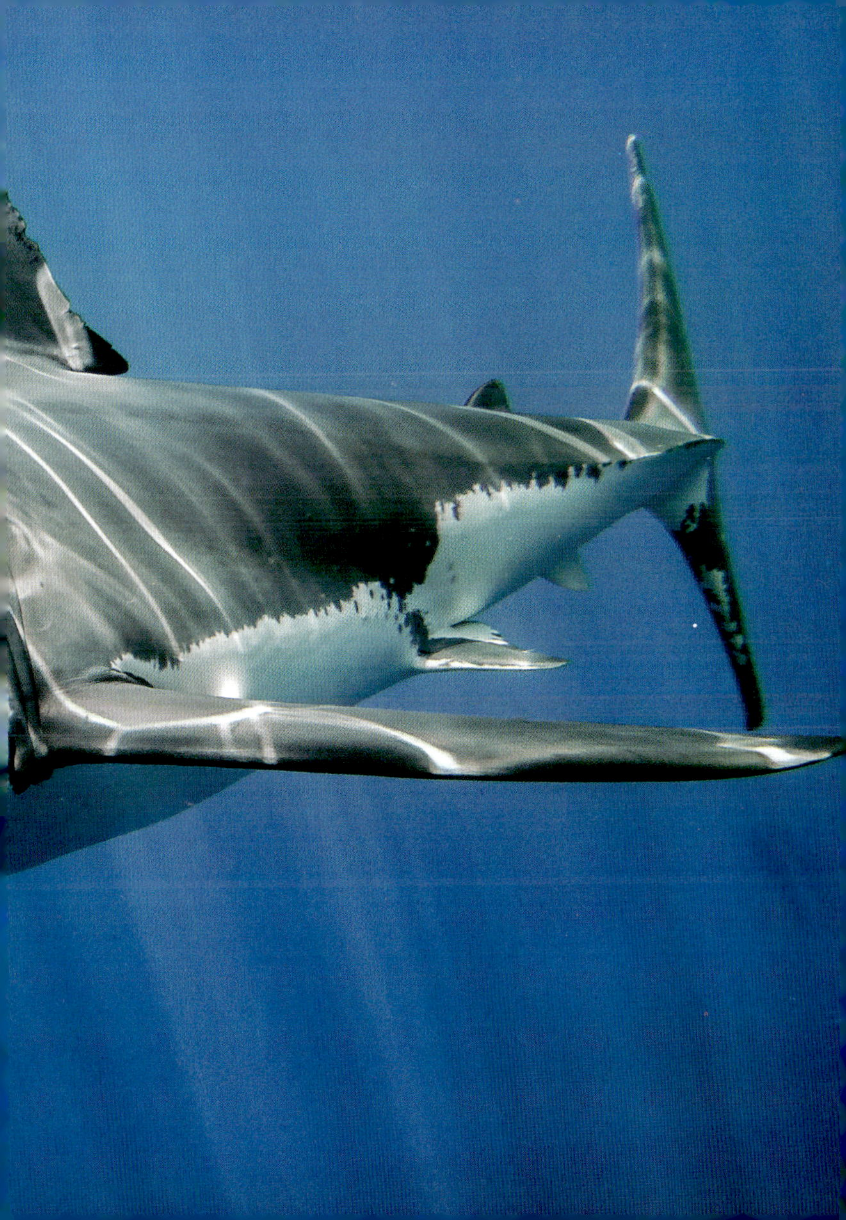

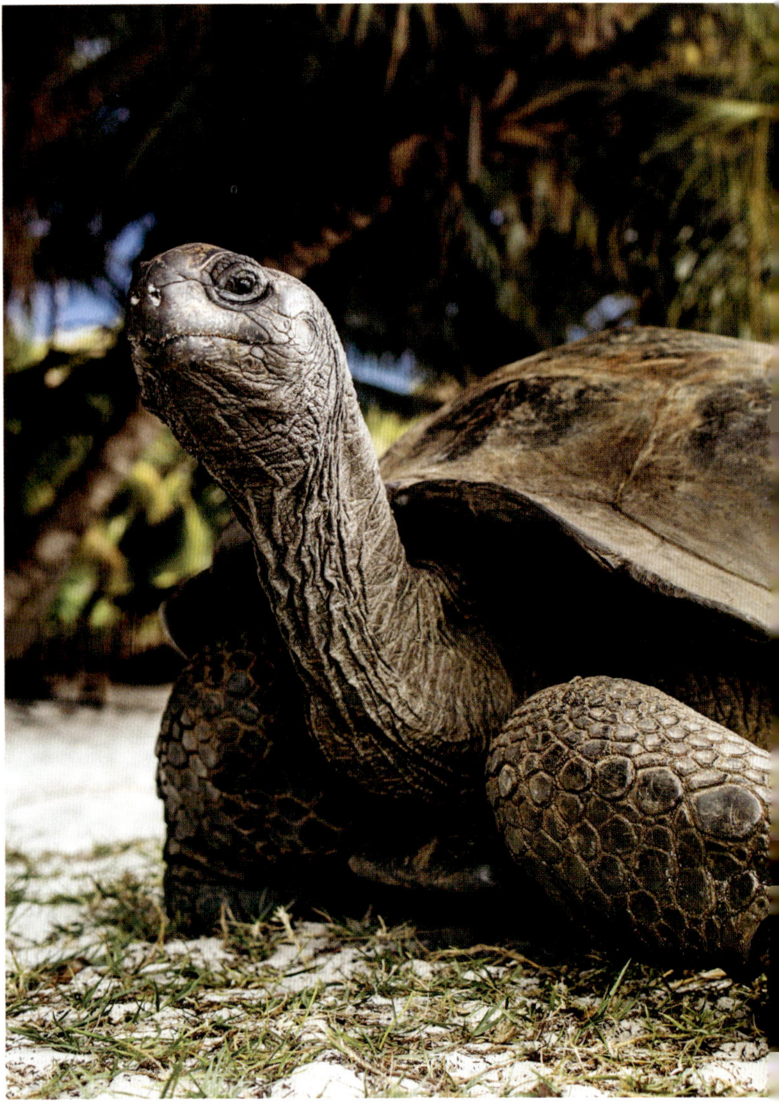

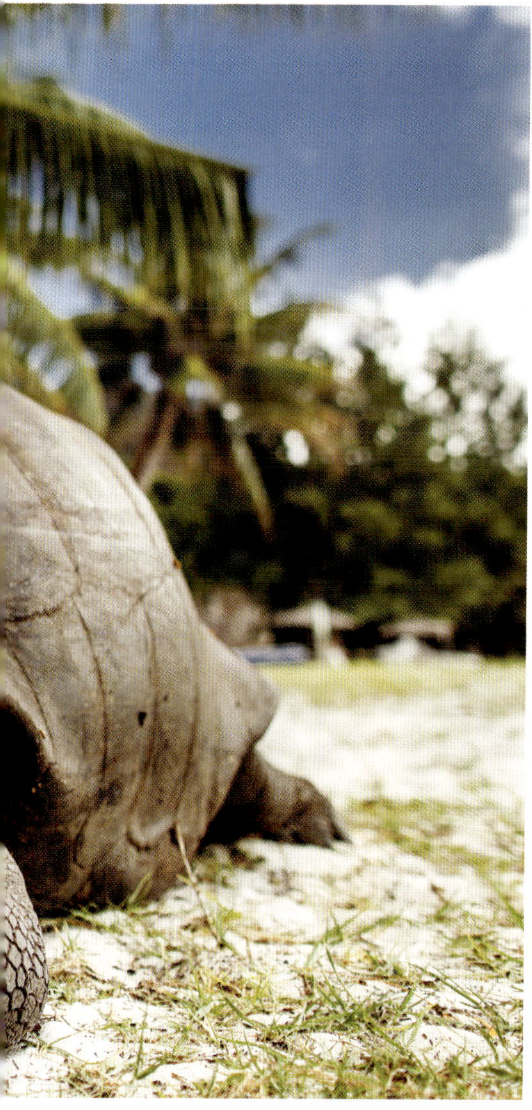

Giant tortoise

The Galápagos in the Pacific and Seychelles in the Indian Ocean, two island groups far from land, are both the last refuges of giant tortoises. In the past, these big reptiles lived on the mainland, but they now only survive in small and highly endangered island populations.

Right whale

Named by whalers for the way they conveniently float once dead, the two Arctic species of right whale are highly endangered, still to recover from their systematic slaughter in the 19th and 20th centuries.

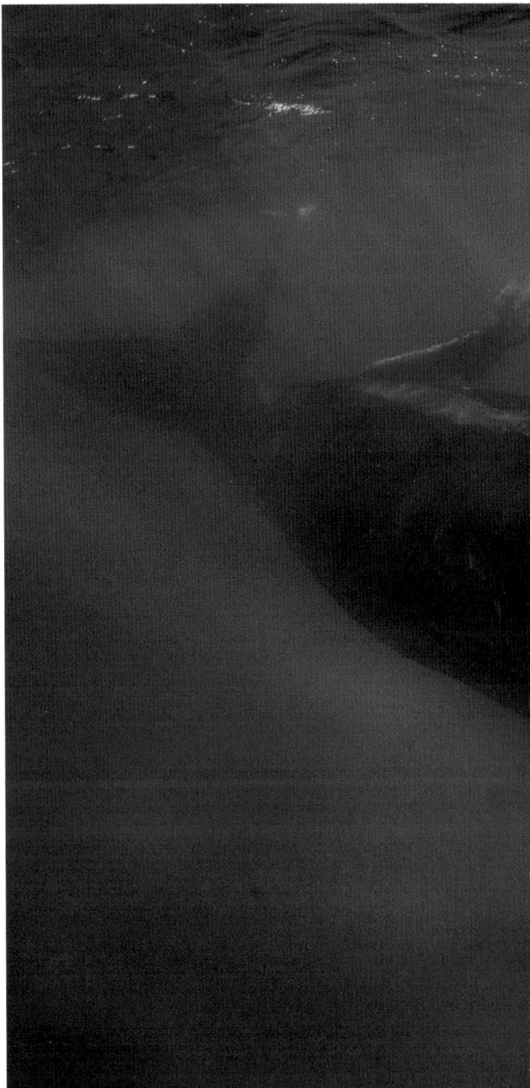

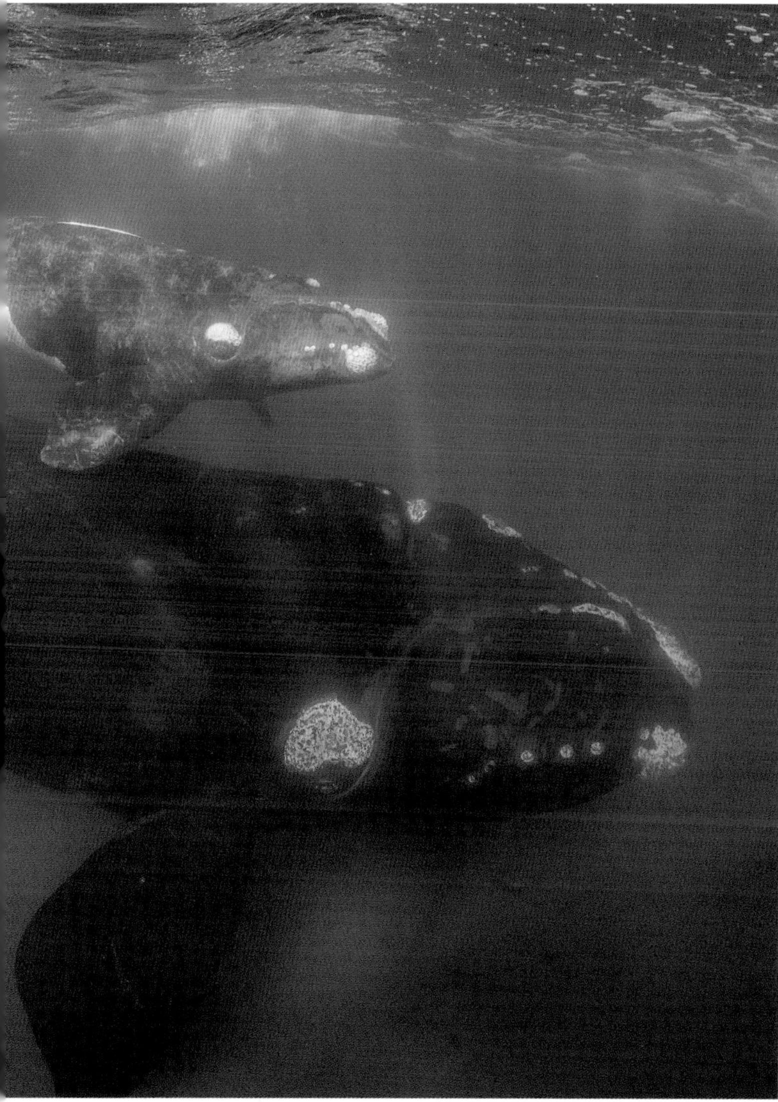

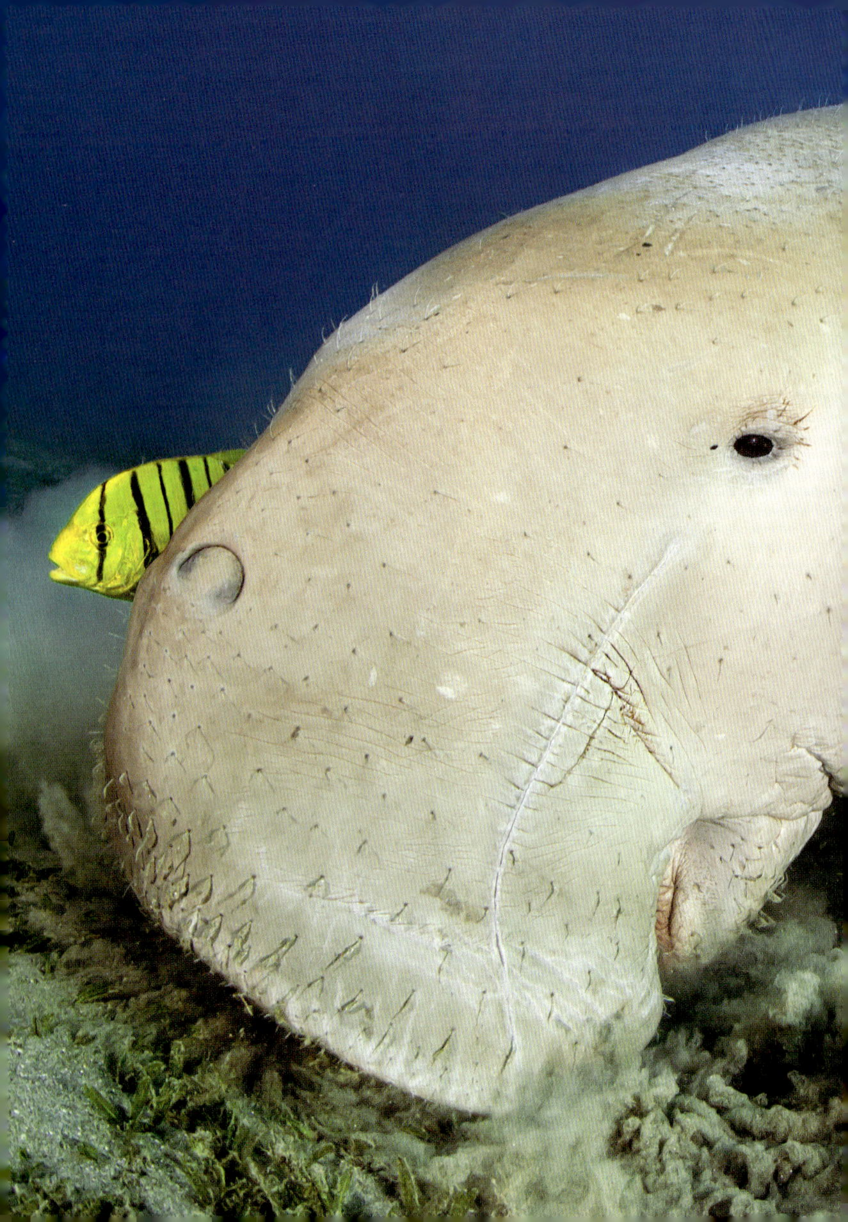

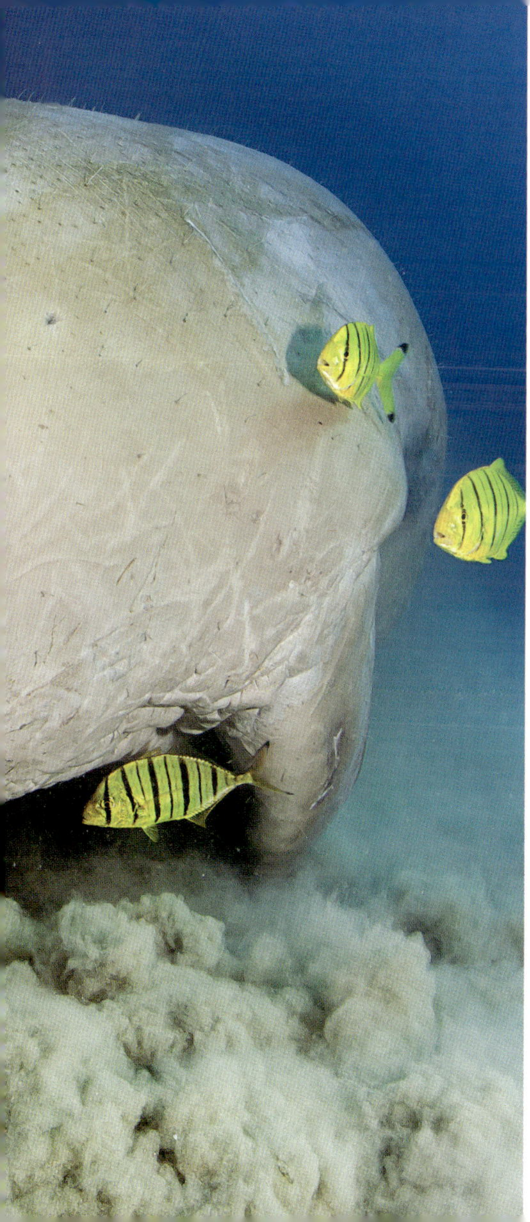

Dugong

The only sea cow species in the Indian Ocean, the dugong is threatened by pollution that kills their feeding sites in warm coastal areas. They also get entangled in fishing nets and drown.

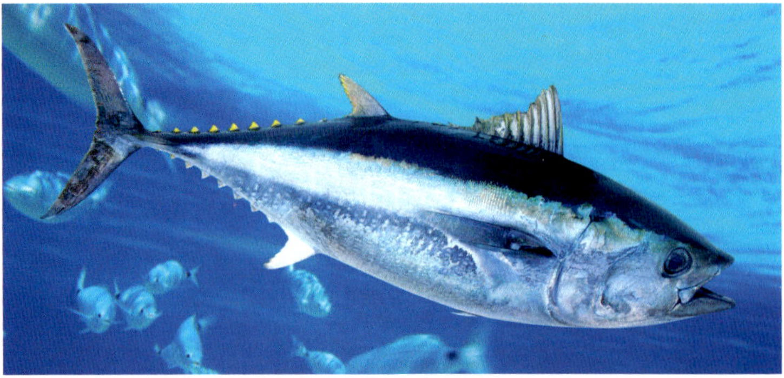

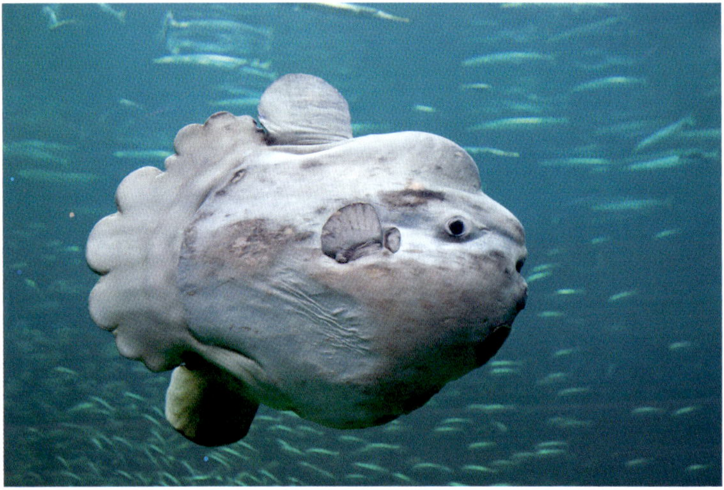

ABOVE TOP:
Southern bluefin tuna
With meat highly prized in East Asia, the bluefin tuna of the Southern Hemisphere is critically endangered, hunted to near extinction. Most authorities ban fishing of the southern bluefin. Stocks are estimated to be at a tenth of normal levels.

ABOVE BOTTOM:
Ocean sunfish
The largest bony fish in the sea, the sunfish is named after the way it is often seen near the surface basking in the sunshine. They can grow to 1.8m (6ft) long and 2.5m (8.2ft) tall! Their decline from being killed in fishing nets is compounded by a slow reproduction rate.

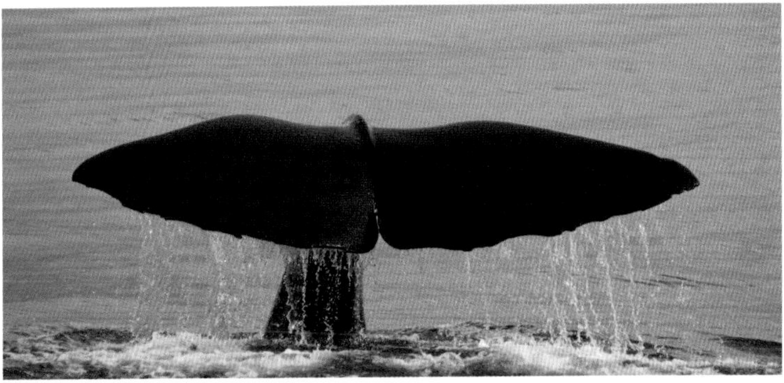

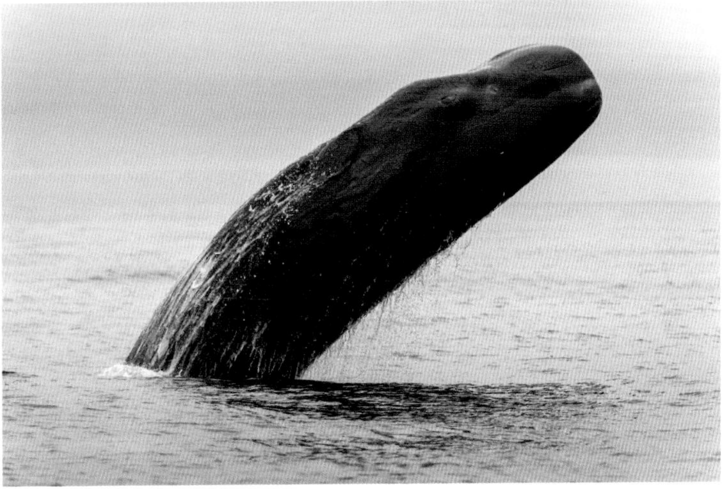

ALL PHOTOGRAPHS THIS PAGE AND OVERLEAF:

Sperm whale

The largest toothed whale, and by the same token Earth's biggest hunter, the sperm whale dives to incredible depths to hunt giant squids in the dark waters. They were hunted relentlessly for their oil until the 1980s, with 750,000 killed during the 20th century. Now left alone, numbers are slowly recovering.

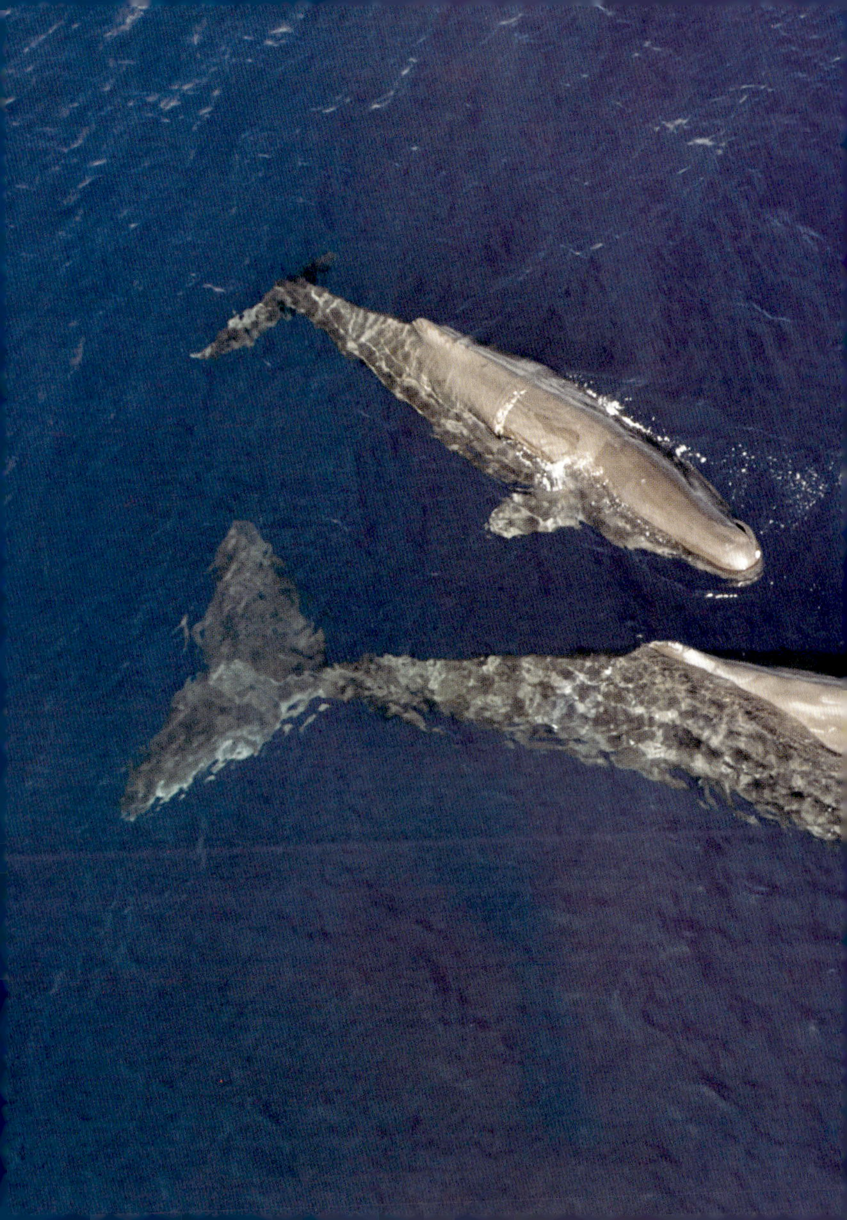

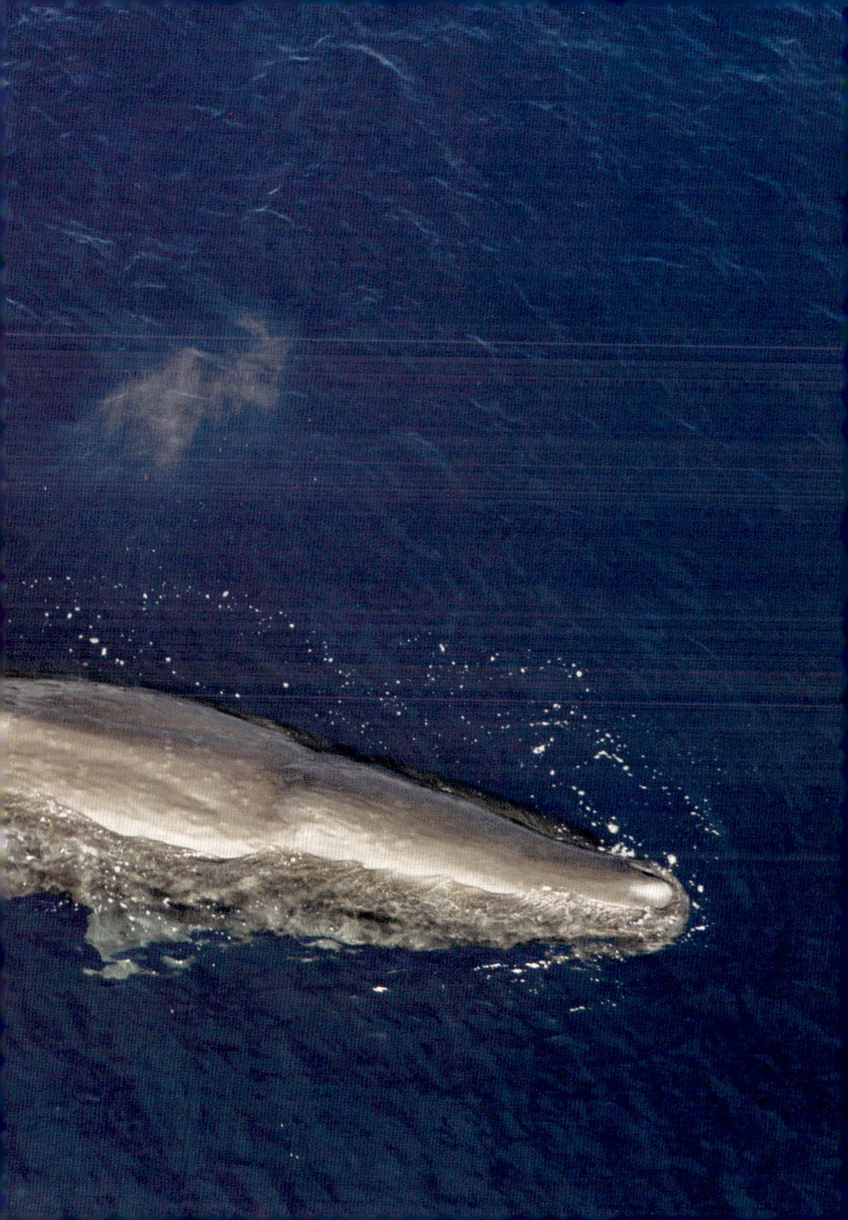

Marine iguana

No other reptile finds all of its food in seawater as does this chunky lizard from the Galápagos Islands. Naturally isolated from the rest of the world, this lizard is under threat from introduced predators like cats and dogs.

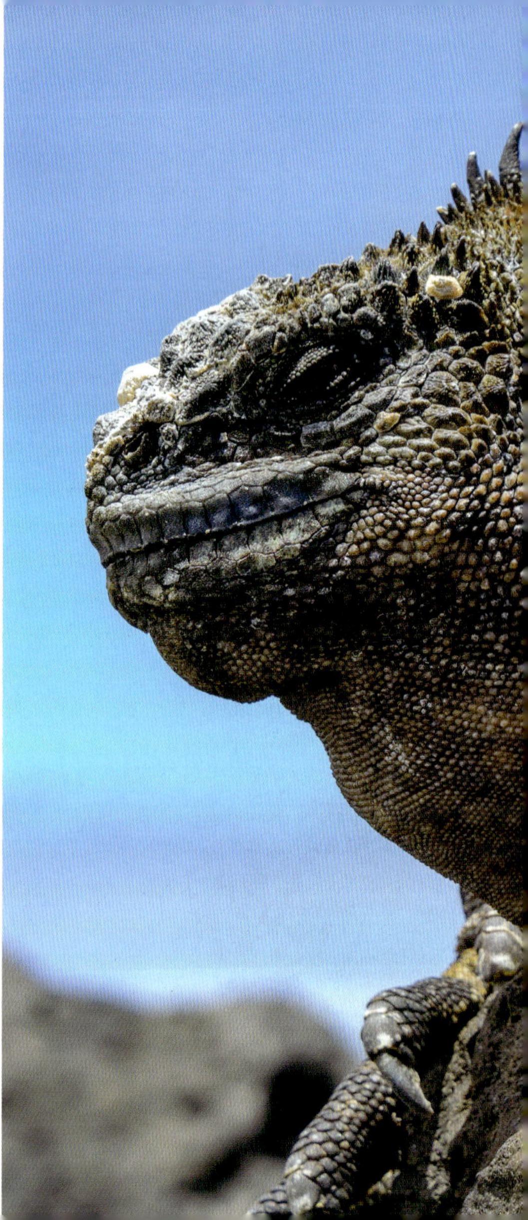

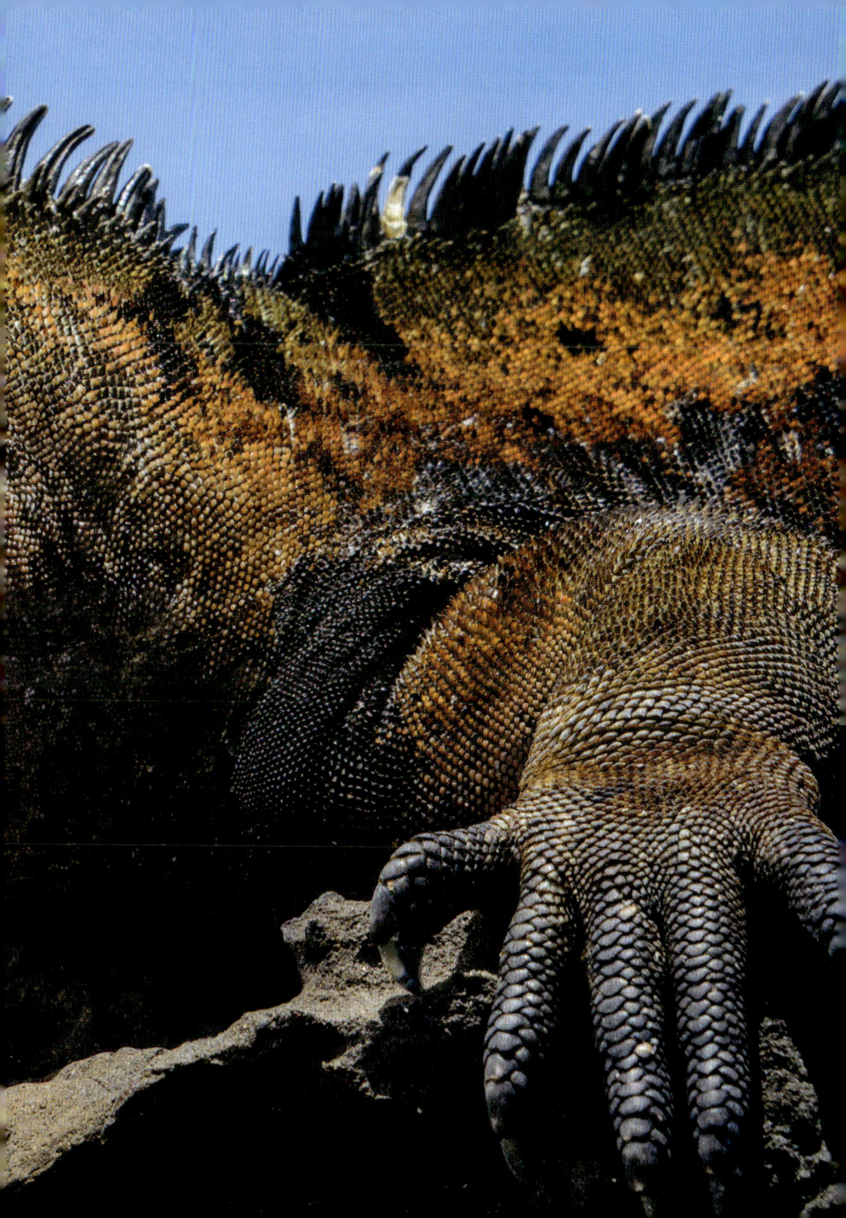

Whale shark

The largest fish in the sea, this shark is a harmless filter-feeder with tiny teeth. Protected around the world, this gentle giant is still hunted illegally to feed the shark-fin market of East Asia.

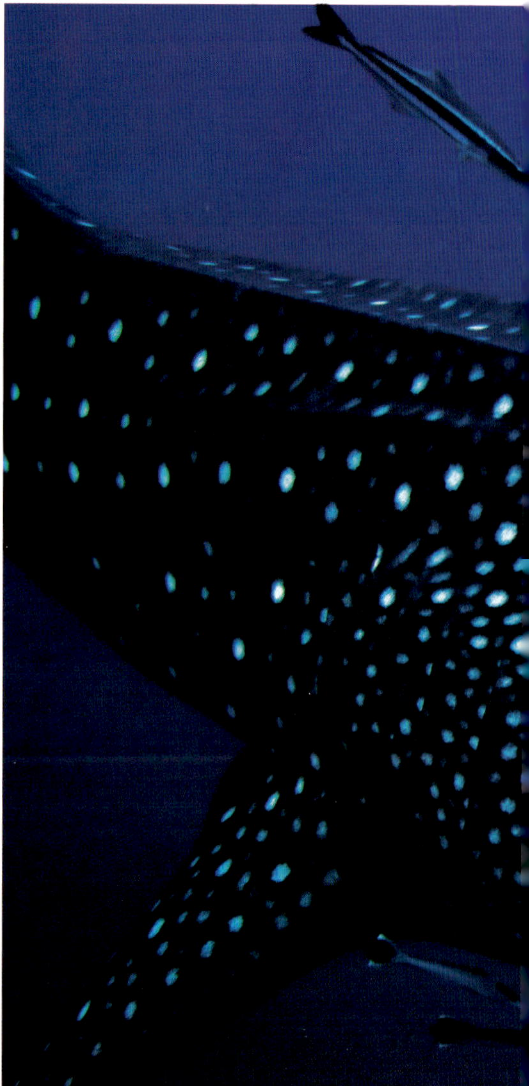

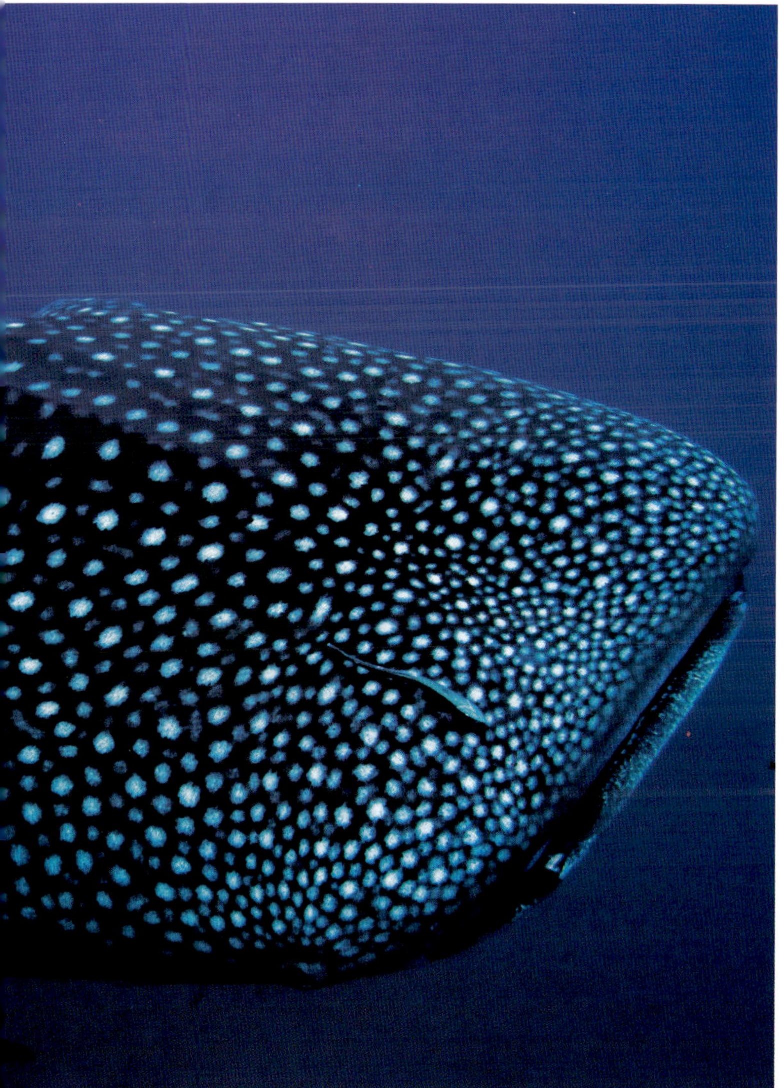

Picture Credits

Alamy: 10 bottom (David Noton Photography), 11 bottom (Suzuki Kaku), 12/13 (Itsik Marom), 14/15 (Zute Lightfoot), 15 (Ariadne Van Zandbergen), 20 (blickwinkel), 20/21 (FLPA), 29 bottom (Ger Bosma), 30/31 (RGB Ventures), 34 (Minden Pictures), 35 (Minden Pictures), 36 top (Nature Picture Library), 42 top (Minden Pictures), 44 (Ger Bosma), 45 (Nature Picture Library), 47 (Nature Picture Library), 58/59 (Johner Images), 62/63 (robertharding), 64 (Louise Heusinkveld), 65 top (Kamal Bhatt), 65 bottom (imageBroker), 66/67 (Sanjay Shrishrimal), 69 (Nature Picture Library), 72 (Xinhua), 76 (blickwinkel), 82 (F1online digitale Bildagentur GmbH), 87 top (Premaphotos), 88 (Reuters), 89 bottom (Auscape International Pty Ltd), 93 bottom (BiosPhoto), 96 (robertharding), 102/103 (Ivana Tacikova), 110/11 (Gerard Lacz), 114/115 (Terry Whittaker Wildlife), 116 top (Luna), 118/119 (Ashish Parmar), 122 (Wolfgang Kaehler), 124 (Kevin Schafer), 134/135 (Juniors Bildarchiv GmbH), 136 top (Kip Evans), 136 bottom (Zoltan Bagosi), 140/141 (Avalon.red), 143 (robertharding), 144 (franzfoto.com), 145 (BiosPhoto), 148 bottom (blickwinkel), 149 (Panoramic Images), 150 (Petra Wegner), 152/153 (Don Johnston BI), 154 (Samantha Crimmin), 155 top (mike lane), 162 bottom (Trevor Baker), 172/173 (F1online digitale Bildagentur GmbH), 183 bottom (Evan Bowen-Jones), 187 top (blickwinkel), 187 bottom (Joe Blossom), 190 (Nature Picture Library), 192 (Matthew Cuda), 195 bottom (Paulette Sinclair), 198/199 (Kirk Hewlett), 204 bottom (BiosPhoto), 205 (Scubazoo), 216 bottom (imageBroker)

Creative Commons Attribution 2.0 Generic: 126 bottom (Ryan Moehring/USFWS)

Dreamstime: 8 (Ecophoto), 16/17 (Fishcat007), 18/19 (illyambradberry), 26/27 (Byrdyak), 28 (Brbraye), 29 top (Lukyslukys), 38 top (Gudkov2000), 38 bottom (Surz01), 39 (Jeep2499), 42 bottom (Jerrymouse), 43 (Anolis01), 48/49 & 49 top (Gudkov2000), 50 top (Duskbabe), 50 bottom (Rhallam), 50/51 (Tatisol), 52/53 (Paulbanton72), 54 (Liumangtiger), 57 (Minyun9260), 58 (Jirousek), 62 top (Sushil1787), 68 (Johncarnemolla), 70/71 (Tee555), 73 (Slowmotiongli), 74/75 (Sean419), 80 (Sourabhis), 81 both (Lukyslukys), 83 (Wlq5231), 84/85 (Ondrejprosicky), 87 bottom (Whitepointer), 89 top (Ken Griffiths), 92 (Clicktrick), 93 top (Cabinpress), 94/95 (Lukyslukys), 98/99 (Kobchaima), 100/101 (Yantar), 106/107 (Mgkuijpers), 108 (Ondrejprosicky), 109 (Thecult), 125 (Wladymyrsergeev), 126 top (Paulreevesphotography), 128 (AmaniImages), 130/131 (Cmoswitzer), 131 bottom (Mattscott), 132/133 (Maloff2), 137 (Gudkov2000), 138/139 (Almartinojeda), 142 (Martinmark), 146/147 (Misad), 155 bottom (Bigm23), 157 (Cfvision09), 162 top (Castenoid), 163 (Matt Cuda), 164 top (Contact93761), 165 (Contact93761), 166 (Anothergecko), 168/169 (Misad), 170/171 (Plowe1), 174 (Kyslynskyy), 176/177 (Ondrejprosicky), 180 (Rostovdriver), 183 top (Bereta), 184/185 (Mrallen), 186 (Mikelane45), 188 (StockPhotoAstur), 189 (Burnel1), 191 (Izanbar), 194 (Chasedekker), 195 top (Flinster007), 200 (Colette6), 202 (Zastavkin), 203 top (Chasedekker), 203 bottom (Jeanro), 207 (Richcareyzim), 208/209 (Ramoncarretero), 214/215 (Berndneeser), 217 top (Ruthie44), 220/221 (Albertoloyo), 222/223 (Djmattaar)

Getty Images: 151 (De Agostini)

Shutterstock: 6 (Johan Swanepoel), 7 (Sergey Uryadnikov), 10/11 (Heinrich Neumeyer), 19 top (Paul Hampton), 19 bottom (Ondrej Prosicky), 22 top (charles Hopkins), 22 bottom (Seyms Brugger), 23 (William Steel), 24 (Claudia Lothering), 25 (Villiers Steyn), 32/33 (Travel Stock), 36 bottom (David Havel), 37 (Sergey Uryadnikov), 40/41 (CherylRamalho), 46 (Danny Iacob), 49 bottom (iPics), 56 top (Dmitry Rukhlenko), 56 bottom (gary718), 60/61 (Nachiketa Bajaj), 62 bottom (Mattiaath), 77 (Charlie Waradee), 78/79 (MehmetO), 86 (NileshShah), 90 top (Kitch Bain), 90 bottom (Boule), 91 (Luke Wait), 97 (Lubos Chlubny), 104 (Eric Gevaert), 105 top (Ondrej Prosicky), 105 bottom (Gudkov Andrey), 112/113 (nuwatphoto), 116/117 (Agami Photo Agency), 117 top (Zuzana Gabrielova), 120 (Anuradha Marwah), 121 top (Michal Ninger), 121 bottom (PhotocechCZ), 127 (Luis D Romero), 129 (SciPhi.tv), 131 top (RealDealRuth), 148 top (FLPA), 156 both (Leonardo Mercon), 158/159 (Helge Zabka), 160/161 (Wayne James), 164 bottom (Sean 7e), 167 (Ronald Wittek), 178/179 (Olga Alper), 180/181 (Nikolai Denisov), 182 (Wirestock Creators), 196/197 (Chase Dekker), 201 (Carmen RL), 204 top (SLSK Photography), 206 (Andrea Izzotti), 210/211 (Pavel Korotkov), 212/213 (wildestanimal), 216 top (lunamarina), 217 bottom (wildestanimal), 218/219 (Marcelo Johan Ogata)